Falling Felines and Fundamental Physics

FALLING FELINES AND FUNDAMENTAL PHYSICS

• • •

GREGORY J. GBUR

Yale
UNIVERSITY PRESS
New Haven and London

Yale University Press books may be purchased in quantity for educational, business, or promotional use. For information, please e-mail sales.press@yale.edu (U.S. office) or sales@yaleup.co.uk (U.K. office).

Set in Adobe Garamond type by Newgen North America, Austin, Texas.
Printed in the United States of America.

Library of Congress Control Number: 2019935202
ISBN: 978-0-300-23129-8 (hardcover : alk. paper)

A catalogue record for this book is available from the British Library.

This paper meets the requirements of ANSI/NISO Z39.48-1992 (Permanence of Paper).

10 9 8 7 6 5 4 3 2 1

Dedicated to my extended cat family:
Sasha, Zoe, Sophie, Cookie, Rascal, Mandarin, Dolly,
Mitzi, Daisy, Hobbes, and the late Simon, Sabrina,
Fluff, Goldie, and Milo

Contents

Preface: Cats Are Crazy

It's fair to say that cats are a bit crazy. As ambush predators that tend to hunt alone, cats have developed an intelligence that allows them to stalk their prey, track it even when it is out of sight, and anticipate its actions. With this intelligence comes a natural playfulness and curiosity that regularly gets cats into trouble. No wonder everyone knows the phrase "curiosity killed the cat."

Fortunately, over countless eons of evolution, cats have also developed important skills that allow them to get out of difficult situations almost as easily as they get into them. Foremost among these is a technique that has been known by many names over the years: cat-turning, the cat-righting reflex, the cat flip, the cat twist. All these names refer to the cat's rather remarkable ability to land on its feet when it falls from a height, no matter what position it was in when it started falling. Cats can do this twist even from small drops of two to three feet by turning in a fraction of a second.

It is a life-saving technique. Cats are often said to have nine lives; if we accept this proverb at face value, I would say that at least four of these lives are due to the cat-righting reflex. Having worked with feline rescue groups, I can attest to the skills of cats. On one occasion, I went to assist in the rescue of a foster cat that had escaped from a home and was high up in a tree. A cherry picker was called so the rescue group could approach the panicked cat and try to coax her from her perch, which was about one hundred feet high. The cat opted to jump, and hit the ground running; when she was brought to the vet, her only injury from the fall was a hairline fracture, which healed.

Cats seem to know that they've got skills, and they like to flaunt them. One member of my extended kitty family, Sophie, used to walk on the *outside* of the railing at the top of the second-floor staircase and ignored all attempts to discourage her. One day my wife happened to see Sophie slip—there was a glimpse of a pair of cat paws, with claws dug into the wood—and then Sophie fell. She landed unhurt but fortunately discouraged from making additional daredevil strolls.

Similar stories of spectacular tumbles abound, and the cat's ability to right itself is common knowledge. What is not broadly recognized, however, is that a significant amount of science is involved in the cat's ability. The physics and physiology of cat-turning have fascinated, frustrated, and baffled scientists for centuries. Although the problem has largely been solved, there are still arguments over the finer details of the cat's ability, and it continues to inspire modern technology.

I first came across the falling cat problem in 2013 while writing my blog Skulls in the Stars, which covers topics from physics to the history of science to weird fiction. I tend to browse old scientific journals for interesting things to write about, and one day I stumbled across the iconic 1894 photographs of a falling cat produced by the French physiologist Étienne-Jules Marey. Intrigued, I wrote a post about Marey and other early falling cat researchers titled "Cat-Turning: The 19th-Century Scientific Cat-Dropping Craze!"

But I wasn't sure that my original explanation for the cat's ability was correct, so I looked for other published research on the falling cat problem. And I kept finding more.

Scientists have been intrigued by falling cats for almost as long as science itself has existed, and this interest has extended across multiple disciplines. Every time one discipline loses interest in the cat problem, another one is right there to find something new to discover.

This book is the story of the falling cat problem, both the science and the history of it. As we will see, falling cats have had a long, remarkable, and sometimes absurd history in the fields of science and engineering. The more that scientists have looked into the problem, the more surprises they have found hidden in the behavior of our furry feline friends. The problem has been connected to some of the most important scientific and technological advances in modern history, from photography to neuroscience to space exploration to robotics and more, and along the way physicists have struggled to explain exactly how cats can do what they do.

This book includes cat photos. Lots of cat photos. Photography has played a huge role in the research on falling cats, so in this book we will look at how photography developed to the point where photographing a falling cat was not only possible but easy. After that, neuroscience picked up the problem and deepened the mystery. Neuroscience research led directly into plans for human space flight, in which falling cats played an oversized role. The combination of neuroscience and physics leads right into the study of robotics, where researchers are still trying to replicate the cat's ability with machines. Along the way, cats have revealed other surprises and caused a lot of mischief in the scientific community.

Scientifically, we have learned a lot from cats, and it's time for that story to be told.

Disclaimer

In 1974, Peter Benchley's novel *Jaws* was published, and its story about a massive killer great white shark was an immediate international sensation, selling some twenty million copies worldwide. The following year, the Steven Spielberg–directed film adaptation of the book was released, becoming the highest-grossing movie of all time, only being dethroned two years later by *Star Wars.*

Whether Benchley or anyone involved with the project anticipated how successful the story would become is unclear. The explosive increase in shark fishing over the next decade was definitely unanticipated, however. The populations of hammerhead sharks, tiger sharks, and great white sharks were decimated, dramatically increasing the threat to the survival of their species. Peter Benchley himself ended up being horrified by this unjustified backlash against sharks and spent the rest of his life advocating for their protection. In a February 23, 2006, interview reported in the *Los Angeles Times,* he said, "Knowing what I know now, I could never write that book today. Sharks don't target human beings, and they certainly don't hold grudges."

I don't expect this book about the history of cat physics to achieve the popular success of *Jaws,* but the novel's unexpected impact on sharks has nevertheless inspired me to make the following request of readers:

Please don't drop your cats!

As we will see through the course of this book, cats as a species are endowed with a remarkable instinct to right their body position when falling, but there are good reasons not to draft a cat into demonstrations:

1. Individual cats may not be very good at it. Though all cats evidently possess the instinct to perform the righting maneuver, not all necessarily do it well, and some might get hurt when dropped.
2. Cats may not enjoy it. I have met a number of cats that have fun with just about any indignity, but not every cat will view falling and righting as a game. Some might object and even hold a grudge against the dropper.
3. Cats may even be traumatized by being dropped. Falling is generally a scary experience for any terrestrial mammal, and it might be genuinely frightening for the average household feline.

The history of falling cat photography stretches back over one hundred years. There are plenty of video resources available online to see how a cat does what it does. The videos generally have the advantage of being in slow motion, so viewers can see all the subtle motions that occur as a cat twists to land right-side up.

Cats as a species have been put through a lot of falls over the past 150 years in the name of science. It is time to give them a well-earned break from active research.

1

Famous Physicists' Fascination with Falling Felines

In the history of nineteenth-century physics, perhaps no name is held in higher esteem than that of James Clerk Maxwell. Born in Scotland in 1831, by the time of his rather early death in 1879 he had made contributions to multiple fields of science and engineering. His greatest achievement was the theoretical unification of electricity and magnetism—thought for thousands of years to be independent forces of nature—into a single fundamental phenomenon, called *electromagnetism*. In the 1860s, Maxwell took a diverse set of observations made by other physicists, distilled them into a complete and self-consistent set of equations, and showed that these equations predicted that electricity and magnetism could combine to form oscillating, traveling electromagnetic waves. Going even further, he brilliantly argued that visible light, long thought to be separate from electricity and magnetism, is in fact an electromagnetic wave.

Maxwell's discovery arguably marks the beginning of the modern era of physics, in which all known physical forces are thought to be manifestations of a single fundamental force; the set of equations that Maxwell completed are now known as *Maxwell's equations* in his honor.

Maxwell also had a reputation for dropping cats.

He earned this peculiar reputation during his university studies, which he started at the University of Edinburgh in 1847 at age sixteen.

James and Katherine Clerk Maxwell, 1869. There is no word on whether James also dropped the dog. Wikimedia Commons.

He moved to Trinity College at Cambridge in 1850, where he studied mathematics and researched the human perception of color. Having distinguished himself as one of the top students, he stayed on at the college as a research fellow for two years. It was during his tenure at Trinity that he spent some of his idle hours investigating how, exactly, a dropped cat can seemingly always land on its feet.

In 1870, Maxwell returned to his alma mater to find that the stories of his cat experiments had grown in his absence. In a letter to his wife, Katherine Mary Clerk Maxwell, he explained the situation: "There is a tradition in Trinity that when I was here I discovered a method of throwing a cat so as not to light on its feet, and that I used to throw cats out of windows. I had to explain that the proper object of research was to find how quick the cat would turn round, and that the proper method was to let the cat drop on a table or bed from about two inches, and that even then the cat lights on her feet."[1] Maxwell seems apologetic in his letter to Katherine, and reassures her that no cats were harmed. Peculiar though his experiment was, it is still striking that it became a legend in just twenty years.

Maxwell was not the only famous scientist of his era to have an interest in falling cats. The Irish physicist and mathematician George Gabriel Stokes (1819–1903) conducted his own informal investigations around the same time. Stokes, like his friend Maxwell, distinguished himself at an early age, earning the coveted position of Lucasian Professor of Mathematics in 1849; he held it until his death. Other holders of the title include the black hole maverick Stephen Hawking, quantum physicist Paul Dirac, computing pioneer Charles Babbage, and the "Father of Modern Physics" himself, Isaac Newton. Stokes certainly deserved to be placed in such heady company, for over his long career he made major contributions to mathematics, fluid dynamics, and optics. All mathematicians and physicists are familiar with Stokes's theorem, which has found application in literally all branches of physics. Stokes's name is also attached to the Navier-Stokes equations, important mathematical formulas used to describe fluid flow (whose properties are still not completely understood). And Stokes made the discovery that fluorescence, the glowing of objects under a black light, involves the conversion of invisible ultraviolet light into visible light.

To this strong scientific pedigree, Stokes added some informal studies of how cats land on their feet. He evidently left no record of his experiments, but his daughter wrote about them in a memoir several years after his death.

> He was much interested, as also was Prof. Clerk Maxwell about the same time, in cat-turning, a word invented to describe the way in which a cat manages to fall upon her feet if you hold her by the four feet and drop her, back downwards, close to the floor. The cats' eyes were made use of, too, for examination by the ophthalmoscope, as well as those of my dog Pearl: but Pearl's interest never equalled that of Professor Clerk Maxwell's dog, who seemed positively to enjoy having his eyes examined by his master.[2]

It is striking that two prominent physicists would be intrigued by a phenomenon so seemingly mundane as a falling cat. What could those two brilliant minds see in the falling cat problem that so many others did not? They saw a secret.

Cats have long been viewed as magical keepers of secrets; in the falling cat problem, we will see how accurate this assessment is.

> Sphinx of my quiet hearth! who deignst to dwell
> Friend of my toil, companion of mine ease,
> Thine is the lore of Ra and Rameses;
> That men forget dost thou remember well,
> Beholden still in blinking reveries,
> With sombre sea-green gaze inscrutable.[3]

2

The (Solved?) Puzzle of the Falling Cat

If Maxwell and Stokes saw something interesting and unusual in the cat-turning problem, they were in a very small minority. As the literature of their era shows, most people found the problem a rather trivial one, one that had already been adequately explained. The conventional explanation was, however, incorrect, and this mistake apparently held up a serious investigation of the cat's righting ability for nearly two hundred years. The wrong argument is intimately tied to the beginning of physics as a formal science.

In the latter half of the nineteenth century, explanations of cat-turning did not appear in scientific journals but in books about cats written by cat lovers. Many such books appeared then, pushing back against the common negative perceptions of felines. For centuries, superstition and ignorance of cat psychology had led western Europeans to dislike cats. Many of those beliefs hold to this day. Cats were—and often still are—viewed as selfish, unemotional, and uncaring about the humans who house them. Cats became a regular fixture in stories of witchcraft, especially black cats; violence against cats was considered acceptable, even reasonable, and people who defended cats were often roundly mocked. As Charles (Chas.) H. Ross lamented in the introduction to his 1893 *Book of Cats*,

> One day, ever so long ago, it struck me that I should like to try and write a book about Cats. I mentioned the idea to some of my friends: the first burst out laughing at the end of my opening sentence, so I refrained from entering into further details. The second said there were a hundred books about Cats already. The third said, "Nobody would read it," and added, "Besides, what do you know of the subject?" and before I had time to begin to tell him, said he expected it was very little. "Why not Dogs?" asked one friend of mine, hitting upon the notion as though by inspiration. "Or Horses," said some one else; "or Pigs; or, look here, this is the finest notion of all:—
>
> 'THE BOOK OF DONKIES,
> 'BY ONE OF THE FAMILY!' "[1]

In spite of this societal condescension, Ross and many others championed cats as pets, friends, and objects of fascination. One cat advocate simply didn't care how others might view his writing. William Gordon Stables, born in Banffshire, Scotland, around 1840, had a life of adventure and independence.[2] While still a nineteen-year-old medical student at Marischal College in Aberdeen, Scotland, he joined a voyage to the Arctic on a Greenland whaling ship, and this was just the beginning of his travels. After graduating as a doctor of medicine and master of surgery in 1862, he earned a commission as an assistant surgeon in the Royal Navy, serving on the HMS *Narcissus* out of the Cape of Good Hope and then on the HMS *Penguin*, which hunted down slave ships off the coast of Mozambique. After serving a couple of years on his Africa assignments, he spent several more stationed in the Mediterranean and the United Kingdom. Health issues made him leave the navy in 1871, but this did not stop his wanderings: he joined the merchant service for an additional two years, sailing around the coast of South America to Africa, India, and the South Seas. In 1875 he settled down at last in Twyford, England, and began a ridiculously prolific writing career, publishing over 130 books. Most were adventure novels for boys, which drew on Stables's own experiences, but he also wrote a number of books on animals and animal care.

Stables's best-remembered work today is probably his guide *"Cats": Their Points and Characteristics, with Curiosities of Cat Life, and a Chapter on Feline Ailments*, which first appeared in print around 1875. The book is a broad survey of everything feline: funny and horrifying cat anecdotes, a discussion of the origins of domestic cats, a guide to feline ailments, advice on teaching cats tricks, arguments in favor of British anticruelty laws for cats, and, more relevant here, an explanation of the cat's ability to always land on its feet.

> Why do cats always fall on their feet? This question is by no means difficult to answer. When she first falls from a height, her back is lowermost, and she is bent in a semicircle. If she fell thus, fracture of the spine, and death, would be the inevitable result. But natural instinct induces her, after she has fallen a foot or two, to suddenly extend the muscles of her back, and stretch her legs; the belly now becomes the convexity, and the back concave, thus altering the centre of gravity, and bringing her round; then she has only to hold herself in this position in order to alight on her feet.[3]

This explanation sounds reasonable and was apparently satisfying to most curious people of the nineteenth century.

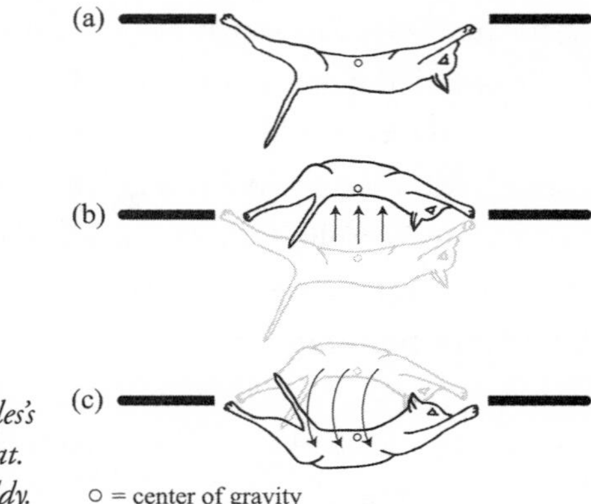

William Gordon Stables's model of the falling cat. Drawing by Sarah Addy.

Imagine a cat hanging by its front and rear paws from two fixed supports; it looks much like the hinged handle of a dresser drawer—(a) in the accompanying figure. Its center of gravity—the point at which gravity effectively pulls on the cat as a whole—is below the supports. When the cat arches its back, as in (b), its center of gravity moves above the supports. This is an unstable position. As long as it keeps its back arched, any small disturbance will make it swing back down so that the center of gravity is below the supports again, as in (c). The cat, which was upside down, is now right-side up!

Stables's argument is simple, compelling, and physically plausible—but wrong. It applies only when a cat is hanging from fixed points, as illustrated, allowing it to move its center of gravity above or below those points. A cat in freefall is not hanging from anything; a change in the cat's body position does not affect its stability at all.

Stables seems to think that the explanation is obvious. He might have first learned it himself from physicist James Clerk Maxwell, whom we've already met. Maxwell earned his first university professorship at Marischal in 1856, only a year before Stables began his medical studies there. Apparently the two interacted, and the young Maxwell left a strong impression on Stables. In his semi-autobiographical novel *From Ploughshare to Pulpit: A Tale of the Battle of Life*, published in 1895, Stables tells the story of a young man who moves up from a farming life to attend Marischal College. Among the descriptions of the various professors, we find the following about Maxwell, without even a change of name.

> Then there was poor Maxwell, so well-known in the scientific world—brown haired, handsome, thoughtful, and wise; he always had some scientific marvel to tell his students during breakfast. He was always smiling, but never laughed a deal. I suppose he had an idea that strong tea was not good for young fellows, for he invariably filled the cup half up with rich delicious cream before pouring in the beverage. Poor Maxwell! He is dead and gone, and [a] great loss his death has been to the world.[4]

Other books, many that predate Maxwell's interest in the feline flip, contain similar explanations of the phenomenon. For example, *First Lessons of Natural History: Domestic Animals*, by M. Battelle, which appeared in 1836, contains this explanation.

> One always sees, with astonishment, that a Cat, falling from a very high point, is always found on his feet, though it seemed to fall on his back. It is not uncommon for a Cat, thrown from the highest level of a high house, to fall so lightly, that it begins to run at the very moment of its fall. This singular effect depends on the fact that, at the moment of falling, these animals bend their bodies, and make a movement as if

> to withdraw: a sort of half-turn results, which makes them fall on their feet, which almost always saves their lives.[5]

Though a description of the center of mass is missing, the explanation is unmistakably the same one that Stables used. But this explanation is even older. It was already being offered in a book of physics exam problems by J. F. Defieu in 1758.

> Question 94: A cat thrown from the third floor into the street has the four legs above, in the first instant of the fall, and falls on all four legs without injuring herself. Why?
>
> Response: The cat, suddenly seized by a type of natural fear, bends the spine of the back, advances the stomach, and lengthens the feet and head as if trying to regain the place whence it comes, which gives the feet and head a greater leverage. In this extraordinary movement, the center of gravity rises above the center of the figure, but, being not sustained, soon descends. As the center of gravity descends, it turns the cat's belly, head, and paws to the ground. Thus the cat, at the end of its fall, finds itself on the ground on its four legs and is, in short, only more vain.[6]

The ultimate clue to the origin of Stables's explanation comes from an 1842 French *Dictionary of Etymology, History, and Anecdotes of Proverbs*. An entry there reads: "It is like the cat, which always falls on its paws."[7] The book of proverbs gives a name to the original author of the cat-righting explanation: Antoine Parent, a largely forgotten French mathematician who published the world's first physical explanation of the cat mystery in 1700.

Antoine Parent was born in Paris in 1666, and he established himself as a mathematical prodigy at an early age. When he was three, he went off to the country to live with his uncle Antoine Mallet, a parish priest, who was known as a good theologian and a talented naturalist. Mallet found the young Parent insatiably curious about mathematics, so he provided the child with all the books on the subject that he could find. Parent studied them and managed to figure out for himself how to perform many mathematical proofs. By age thirteen he had filled the margins of numerous books with annotations and commentary.

Not long afterward, he was apprenticed to a family friend who taught rhetoric in the city of Chartres. This teacher possessed in his room a model that illustrated how sundials need to be designed differently at different locations upon the Earth. This model was in the form of a dodecahedron, a symmetric twelve-sided geometric solid. Upon each face of it was marked a sundial appropriate for its relative position upon the Earth. Parent, fascinated by the subtlety of the sundial design, attempted to deduce the underlying mathematics on his own. He failed—not surprising at age fourteen—but his master explained to him how the proper construction of sundials depends upon the underlying spherical geometry of the Earth. The intrepid Parent then undertook to write his own amateur book on the art of making sundials, or gnomonics.

Though mathematics was his passion, Parent suffered the fate of many brilliant artists and scientists: he was persuaded by his friends to travel to Paris to study to be a lawyer, for law was a more lucrative profession than math (back then, as now). The moment he finished his law degree, however, he shut himself away in a residence at the College of Dormans-Beauvais in Paris, where, living on a poverty-level income, he dedicated himself to the study of mathematics. His only excursions were to visit the Royal College of Paris to interact with—and hear lectures by—prominent academics, such as the mathematician Joseph Sauveur, who studied geometry and the science of sound.

Parent was an enterprising sort, and the outbreak of the Nine Years' War, between France and a coalition of European countries, gave him the opportunity to supplement his income with teaching. Since the conflict had led to a high demand for soldiers and scholars who could understand the mathematics and engineering of warfare, Parent took on students to teach them the theory of building defensive fortifications. In this, Parent had no personal experience; he undoubtedly drew on a number of old texts on the subject, such as Jean Errard's 1600 volume, *La Fortification Démonstrée et Réduicte en Art*.[8]

But Parent began to have scruples about teaching a subject with which he had no practical knowledge. He explained his concerns to Joseph Sauveur, and Sauveur offered assistance in the form of a job recommendation. With his help, Parent was introduced to Marquis d'Aligre, a nobleman campaigning in the Nine Years' War who happened to need the services of a mathematician. Parent worked under d'Aligre for two military campaigns, which earned him a reputation as a brilliant scientist, mathematician, and thinker.

In 1699, Parent was able to use this reputation to his great benefit. That year the mathematician Gilles Filleau des Billettes was admitted to the Royal Academy of Sciences in Paris in a distinguished role as a "mechanician," and he took in the highly regarded Parent as his disciple. Parent, now with a stable academic position, indulged in a dizzyingly diverse range of studies, including anatomy, botany, chemistry, mathematics, and physics. But his lifelong enthusiasm and freewheeling nature now began to work against him.

> But this extent of knowledge, joined to a natural warmth and impetuosity of temper, raised a spirit of contradiction in him, which he indulged on all occasions; sometimes to a degree of precipitancy that was highly culpable, and often with but little regard to decency. Indeed the same behaviour was returned to him, and the papers which he brought to the academy were often treated with much severity. In his productions, he was charged with obscurity; a fault for which he was indeed so notorious, that he perceived it himself, and could not avoid correcting it.

This excerpt, which comes from a 1795 mathematical dictionary, is itself essentially an English translation of the memorial written for Parent by his fellow members of the Paris Academy after his death.[9] His colleagues found his attitude so obnoxious that they recorded it for posterity, in his obituary no less.

Parent nevertheless regularly reported his results to the Academy, which published accounts of them in the society's journal, *Histoire de l'Academie Royale*, up until his death. One of the very first, which would provide almost 200 years of (incorrect) explanations of the falling cat problem, was printed in 1700. It is titled "Sur les corps qui nagent dans des liqueurs," which may be translated as "On the Bodies That Swim in Liquids."[10] At first glance, this would appear to be completely unrelated to the physics of falling cats—but appearances can be deceiving, especially where the eclectic Parent is concerned.

Parent's paper concerns the buoyancy of objects that are submerged in water. In 250 BCE, the Greek philosopher and mathematician Archimedes was the first to state that the buoyant force that lifts a submerged object is equal to the weight of water displaced by that object. There are, therefore, two forces acting on any submerged

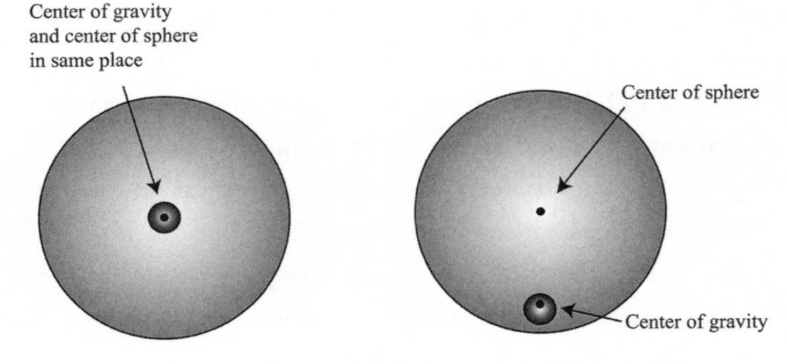

Dynamics of spheres suspended in fluid. My drawing.

object: the force of gravity, pulling the object down, and the buoyant force, pushing it up. If the object is heavier than the water it displaces, then it will sink; if it is lighter than the water it displaces, then it will float.

The density of water in a deep pool increases with depth, which means that the weight of a fixed volume of water is greater the deeper one goes. A sphere of wood, which is lighter than the equivalent spherical volume of water at the surface of the pool, will therefore be pushed upward and float at the surface; a sphere of lead, which is heavier than an equivalent spherical volume of water at the bottom of the pool, will sink to the bottom. By making a sphere that consists of a small lead core surrounded by a wooden shell, as illustrated in the left part of the accompanying figure, it is possible to tailor the size of the core so that the combined object is suspended at some depth below the surface of the water; it "swims," using the terminology of Parent.

But suppose the sphere is made in an asymmetric way: the lead core is embedded away from the wooden sphere's center, as in the right part of the figure. Now the center of gravity of the sphere is no longer in the sphere's center but closer to the lead core. How does the behavior of this sphere compare to that of the sphere with the central lead core?

This question had already been considered by the Italian physicist Giovanni Alfonso Borelli several decades earlier, in his two-volume 1685 text *De Motu Animalium*, "The Motion of Animals." Borelli was interested in studying the various movements of animals and their constituent muscles through the use of mathematics and physics. Because of his important investigations in this subject, and his strong view that animals could be viewed as complex machines, Borelli is today often regarded as the "Father of Biomechanics."

Borelli's studies of the sphere problem were motivated by an interest in how animals move in water. For the unevenly weighted sphere, Borelli argued that the sphere, when dropped into a body of water with the lead side up, would first sink to the point where the forces of buoyancy and gravity were equal and *then* it would rotate around its center until the center of gravity—the lead core—was lowest.

Parent considered the motion to be somewhat more complicated, and he had the advantage of more sophisticated physics to back up his case. Isaac Newton's seminal book *Philosophiae Naturalis Principia Mathematica* (Principles of Natural Philosophy), often simply called *Principia*, was published in 1687, a couple of years after Borelli's work, and it presented for the first time a unified mathematical theory of the motion of massive bodies. Given some of the insights of Newton's work, Parent noted that the force of gravity and the force of buoyancy would act at different points on the sphere. The buoyant force would pull upward at the geometric center of the sphere, whereas gravity would pull downward at the center of gravity. Because of this application of forces at different locations, Parent argued, the sphere would necessarily rotate around some location between these two centers, and, furthermore, this rotation would happen while the sphere was still descending toward its equilibrium depth.

An object rotating while it falls sounds very much like a falling cat, and Parent clearly thought the same thing. After laying out his mathematical arguments, he noted:

> Hence the cats, and several animals of the same kind, as the martens, polecats, foxes, tigers, etc., when they fall from an elevated place, usually fall upon their paws, though their paws were at first above them, and the animals would consequently have fallen on their heads. It is quite certain that they could not by themselves overturn themselves in the air, where they have no fixed point to support themselves. But the fear which they experience makes them bend the spine of the back, so that their midsection is pushed upward, and at the same time lengthen their heads and legs to the place from which they have fallen as if to find it, which gives these parts greater leverage. Thus their center of gravity comes to be different from the center of the figure, and above it, whence it follows, by the demonstration of M. Parent, that these animals must make a half-turn in the air, turning their legs down, which almost always saves their lives. The finest knowledge of mechanics would not be better on this occasion than that of a confused and blind poetry.

Physicists are infamous for trying to solve problems by simplifying them until they are in a ludicrously unrecognizable form. A long-existing joke among students of the subject involves a physicist modeling a cow by first saying, "Now, for simplicity, imagine that the cow is spherical." In Parent's paper, he has almost literally done this, describing a falling cat as a buoyant sphere.

Parent's explanation is the origin, more or less, of the argument that William Gordon Stables and other cat lovers would repeat over 150 years later: the cat arches its back, pushing its center of gravity above some pivot point, which causes it to swing to an upright position. Stables was vague on what point the cat was supposedly pivoting around; in Parent's original argument, however, we see that the cat supposedly flips around what we could call a point of buoyancy.

The argument is incorrect. It is true that the air has a buoyant force, which, for example, causes helium balloons to float upward when released. But for cats and humans this force is almost negligible compared to the force of gravity: we, as humans, don't find ourselves floating around in our daily lives. A cat flips over in a fraction of a second, and for this to happen through buoyancy, the two opposing forces would have to be of almost equal strength.

It is interesting to note, however, that Parent's strategy more or less works for falling humans, although it is wind resistance, not buoyancy, that causes the flip. A skydiver falling at high speeds experiences a strong upward force of wind resistance; at terminal velocity, the wind resistance perfectly balances the gravitational force. The stable falling position for a skydiver is belly to Earth, with the back arched. The center of gravity of the skydiver is, therefore, at the skydiver's lowest point. If a skydiver accidentally ends up with his back turned toward the Earth in freefall, the standard training is to arch the back, like Parent's cat; the jumper will quickly flip back belly to Earth.[11]

Antoine Parent remained a prolific researcher, publishing numerous books and papers until succumbing to smallpox in 1716. His explanation of the falling cat would survive long after him—in fact, long after Parent himself was remembered.

Having traced the scientific origins of falling cat research back to the year 1700, we might naturally ask whether any explanations predate Parent's work. Certainly people had noticed the cat's remarkable ability long before then, and some of those people must have wondered how a cat can perform such a feat.

At least one other early scientist and philosopher may have studied falling cats, but with a very different motivation from those of the investigators who came later. Most people are familiar with the scientist-philosopher René Descartes (1596–1650), who is famous for his declaration "I think, therefore I am." This statement is an early ontological proof of human existence: "How do I know that anything truly exists? Well, I am able to think and reason and ask the question, so at the very least *I* must exist."

Descartes was a pioneering natural philosopher and mathematician, though his science was often inextricably connected to his religious beliefs. While Descartes was working in the 1630s on his book *La Monde*, his first full account of his natural philosophy, he was also preoccupied with the question of whether animals have souls. In the city of Leiden, where he did this work, there is a legend that Descartes threw a cat out of a first-floor window to see whether the animal exhibited fear, something that, in his view, only a creature with a soul would have.[12]

This legend has a disturbing plausibility to it. Over the same period of time, Descartes performed a lot of horrific animal vivisection experiments, also with the ostensible goal of seeing whether animals experience sensations and emotions as humans do. Clearly, Descartes was of the opinion that they do not. Tossing a cat out of the window could have been the least cruel thing he did, for the cat's

The field is of the Saphire, on a chiefe Pearle, a Musion, or Catte, Gardant, Ermines. This beaste is called a Musion, for that he is enimie to Myse, and Rattes. And he is called a Catte of the Greekes, because he is slye, and wittie: & for that he seeth so sharpely, that he ouercommeth darknes of the nighte, by the shyninge lyghte of his eyne. In shape of body

The coat of arms and a partial description of the "Catte of the Greekes." From Bossewell and Legh, Workes of Armorie, *p. F o.56.*

righting reflex would have almost guaranteed a safe landing. Presumably the cat in question would have gone in search of a more hospitable home as well.

It is possible to find even earlier observations of the falling cat ability, albeit of a non-scientific nature. In the 1572 book *Workes of Armorie,* author John Bossewell catalogs and categorizes the coats of arms of various noble families; most of these coats of arms are based on animals, and the animals' strengths are taken as symbolic of the strengths of the families. Of the "Cattes," Bossewell notes, "He is a cruell beaste, when he is wilde, and falleth on his owne feete from moste highe places: and noneth is hurte therewith."[13] Clearly, the cat-righting ability was held in high enough regard to be considered a worthy heraldic attribute.

Another explanation of the cat-turning ability, albeit non-scientific, possibly predates Bossewell's heraldry description. It is a story about the Prophet Muhammad, founder of Islam, who lived from 570 to 632. If it is even remotely contemporary with the Prophet himself, it would rank as the earliest known explanation of cat-turning.

One version of the story is reproduced here in full:

> The Prophet Mohammed had one day gone far into the desert, and after walking a long distance, fell asleep, overcome with fatigue. A great serpent—may this son of Satan be accursed!—came out of the bushes and approached the Prophet, the Messenger of Allah—whose name be glorified! The serpent was on the point of biting the Servant of the All-Merciful, when a cat, passing by accident, fell upon the reptile, and, after a long struggle, killed it. The hissing of the expiring monster awoke the Prophet, who understood from what danger the cat had saved him.
>
> "Come hither!" commanded the Servant of Allah.
>
> The cat approached, and Mohammed caressed him three times, and three times he blessed him, saying, "May peace be upon thee, O cat!" Then in further token of his gratitude, the Messenger added, "In return for the service thou hast done me, thou shalt be invincible in combat. No living creature shall be able to turn thee on thy back. Go, thou art thrice blessed!"
>
> It is in consequence of this benediction of the Prophet that a cat always alights on its feet from whatever height it may fall.

This version of the story appears in the 1891 book *The Women of Turkey and Their Folk-Lore*, written by English folklorist Lucy Mary Jane Garnett.[14] It in turn references an 1889 French book of world oral traditions; the story there comes directly from a 52-year-old theology student. I have been unable to trace the legend further back, at least in English, so its antiquity remains an intriguing question.

What is undeniable, however, is that cats have historically been treated with much greater respect in the Muslim world than in the Western world, dating back all the way to the Prophet's documented love for felines and continuing to this day, though arguably with less reverence.[15] A wonderful example comes from William Gordon Stables and his book of cats, where he shares the following tale of his travels.

> I had been explaining to the gentleman, that my reason for not being off the night before, was my finding myself on the desert side of the gates of Aden after sundown. A strange motley cut-throat band I had found myself among, too.
>
> . . .
>
> I myself was armed to the teeth: that is, I had nothing but my tongue wherewith to defend myself. I could not help a feeling of insecurity taking possession of me; there seemed to be a screw that wanted tightening somewhere about my neck.
>
> . . .
>
> In the midst of a group of young Arabs, was one that attracted my special attention. He was an old man who looked, with his snow-white beard, his turban and robes, as venerable as one of Dore's patriarchs. In sonorous tones, in his own noble language, he was reading from a book in his lap, while one arm was coiled lovingly round a beautiful long-haired cat. Beside this man I threw myself down. The fierceness of his first glance, which seemed to resent my intrusion, melted into a smile as sweet as a woman's, when I began to stroke and admire his cat. Just the same story all the world

> over,—praise a man's pet and he'll do anything for you; fight for you, or even lend you money. That Arab shared his supper with me.
>
> "Ah! my son," he said, "more than my goods, more than my horse, I love my cat. She comforts me. More than the smoke she soothes me. Allah is great and good; when our first mother and father went out into the mighty desert alone, He gave them two friends to defend and comfort them—the dog and the cat. In the body of the cat He placed the spirit of a gentle woman; in the dog the soul of a brave man. It is true, my son; the book hath it."[16]

In returning to William Gordon Stables, we have come full circle in our historical investigation. Antoine Parent introduced the first physical explanation of cat-turning, which, though incorrect, would last for almost two hundred years. James Clerk Maxwell and George Gabriel Stokes suspected that something interesting remained to be discovered in the cat's trick, but they were unable to make any further progress owing to one simple obstacle: the limits of human vision. Assuming that a cat can turn itself over when falling a distance of two feet, as (we assume) Maxwell said, implies that the cat can turn over in one-third of a second. This is generally too fast for the human eye to discern what, exactly, the cat is doing during its fall.

Fortunately, about the same time that Maxwell and Stokes were being baffled by the falling cat, a new technology developed that would allow researchers to study the cat's motion in freefall in detail. As far as the cat problem was concerned, however, this technology would raise many more questions than it would answer.

3
Horses in Motion

Paintings are often intended to record a moment in history, to capture it for posterity in the way the artist himself or herself perceived it. Sometimes the finished product captures more history than the artist intended, or even imagined.

Among the countless pieces of artwork at the Louvre is an 1821 painting by Théodore Géricault entitled "The Epsom Derby." Géricault is known today as one of the pioneers of Romanticism in art, a movement that emphasizes emotion, as well as glorification of the past and natural settings. His most famous work applying these themes is "The Raft of the Medusa," painted between 1818 and 1819, which shows the survivors of the French naval frigate *Méduse*, desperate and dying, adrift in a merciless sea. The painting, a stylized depiction of a disaster that occurred in 1816, caused quite a bit of controversy at its unveiling. "The Epsom Derby," painted only a few years later, is almost opposite in tone: it depicts four horses in the midst of a race, their jockeys urging them on to victory.

To a modern viewer, the painting looks somewhat odd, though it may take a moment of reflection to determine why. The four horses are in exactly the same pose, implying that they are galloping in perfect unison, and with such extreme extensions of their limbs that they almost appear to be levitating. The bottoms of the horses' rear hooves are pointed upward, while the front hooves reach far forward. Today,

Théodore Géricault, "The 1821 Derby at Epsom," 1821. Wikimedia Commons, The Yorck Project.

most of us have at least an intuitive feeling that horses do not run this way.

Géricault was not the only artist to depict horses in this manner. It was a standard pose among nineteenth-century painters. Similar depictions of what is now called the flying gallop appear in artwork dating back thousands of years. Its strongest historical competitor is a pose in which the horse appears to be leaping, with its front hooves up and the rear hooves on the ground.

The artists did not paint the animals this way from lack of talent. Rather, they were limited by what their eyes could tell them. A single cycle of a galloping horse's legs takes place in a fraction of a second, much too fast for the human eye to see, just as cat-turning was too fast for Maxwell and Stokes to discern. Lacking direct knowledge of the motion of a horse, artists may have used other animals as a

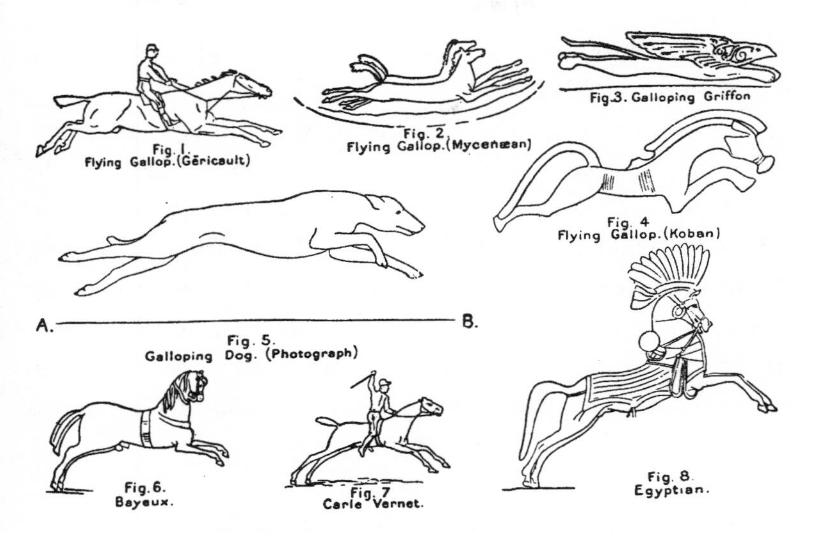

Ray Lankester's "The Problem of the Galloping Horse," plate II, comparing artistic representations of galloping horses through history with a depiction of a running dog. The Mycenaen gallop comes from a dagger circa 1800 BCE. From Lankester, "The Problem of the Galloping Horse," p. 57.

visual analogy. In "The Problem of the Galloping Horse," published in the early 1900s, Sir Ray Lankester postulates that the ancients may have crafted the flying gallop from watching dogs.[1] Dogs run at a much slower pace than horses and, owing to their size, are much easier to see in their entirety in a single field of vision. Their running gait includes a pose that could be considered the dog equivalent of the flying gallop.

For most of history, the study of animal motion was largely limited by the speed of the human eye. This began to change in the mid-1800s as chemistry and optics were unified into the science, technology, and art of photography. The new process would make available the answers to many questions yet simultaneously introduce many new ones, showing among other things that the galloping of a horse and the maneuvers of a falling cat are more sophisticated than anyone at the time imagined.

The key elements needed for the development of photography had been in existence long before the 1800s. One of these was the camera obscura ("dark room"), a box or a chamber closed off to outside light with the exception of a single small hole. The light traveling through the pinhole produces a high-quality, albeit inverted, image of the view outside. This unusual way to form an image has been recognized, on and off, for at least two thousand years. The earliest known written description comes from the Chinese philosopher and scholar Mozi, circa 400 BCE.

> Canon: The turning over of the shadow is because the crisscross has a point from which it is prolonged with the shadow.
>
> Explanation: The light's entry into the curve is like the shooting of arrows from a bow. The entry of that which comes from below is upward, the entry of that which comes from high up is downward. The legs cover the light from below, and therefore form a shadow above; the head covers the light from above, and therefore forms a shadow below. This is because at a certain distance there is a point which coincides with the light; therefore the revolution of the shadow is on the inside.[2]

In other words, light coming from a high point outside the box passes through the pinhole and appears as a low point in the image, and vice versa. Among many others who recognized and studied the imaging properties of the camera obscura are the Greek philosopher Aristotle

(384–322 BCE), the Muslim scholar Ibn-al-Haytham (965–1039), and the Italian polymath Leonardo da Vinci (1452–1519).

In spite of its long history, the technique only became popularized in the late 1500s by the Italian scholar Giovanni Battista della Porta (c. 1535–1615). In his 1558 book, *Natural Magic*, he provided a detailed description of the image-forming properties of the camera obscura and the best way to view the image in a closed room.[3] This book drew the public's attention to the technique, which consequently remained popular for the next few centuries. The camera was viewed primarily as a source of amusement, a way to "magically" make images appear in a darkened room, but artists also saw its potential to allow easy sketching of landscapes. In the *Dictionnaire Technologique* of 1823, for example, we find the following description, in which the camera is referred to as a "darkroom." "The darkroom is of frequent use; not only does it offer recreation by forming animated pictures of a varied and very amusing aspect, when one has a window from which one discovers a beautiful horizon, but also one uses it to draw views and landscapes quickly, or to draw out prospects, which, without this apparatus, would require much time, and which are of extreme fidelity."[4]

For illustrators working with the camera obscura, it would not be a great leap to start imagining how much more elegant the process would be if the images could be recorded automatically, without the intervening labor of the artist.

Chemistry was the other key element needed to make the dream of recording images a reality. Chemists had long known that some materials will react and change color when exposed to light, either blackening or bleaching, and that this change can happen relatively quickly. In 1717, for example, the German scientist and physician Johann Heinrich Schulze discovered that a mixture of chalk, nitric acid, and silver turns black when exposed to light, and he used this reaction to baffle and amuse his friends. He poured the mixture into a bottle and wrapped the bottle in a paper stencil with words cut into it; when exposed to the light, the shapes of the words were blackened in the mixture. A simple stirring of the liquid would then cause the impressions to disappear.[5] Schulze, though, did not use this effect for anything other than entertainment.

Others followed in Schulze's footsteps, noting how patterns could be made in certain chemical mixtures via the use of appropriate stencils. None of these demonstrations count as "photography," however, in that none of the demonstrators used the process to faithfully reproduce a scene as seen by the human eye. The collision of camera and chemical would finally be achieved by a brilliant French inventor, with significant help from his elder brother, in the early 1800s.

Joseph-Nicéphore Niépce was born in 1765 into a wealthy and learned family in Chalon-sur-Saône, in eastern France. His father was a successful lawyer, and the family had for centuries had a high social status due to their wealth and property holdings. Nicéphore therefore had the freedom to develop his natural curiosity and mechanical talents. His brother Claude, Nicéphore's elder by roughly two years, was similarly gifted. It is reported that the two of them took well to the teachings of their clerical tutor and that they built small wooden machines together in their spare time. From a young age, Nicéphore had dreamed of being a priest, and when he finished his education, he taught at a Catholic college.

This professorship might have marked the end of his career as an inventor, but the start of the French Revolution in 1789 pushed him along a different path. The ecclesiastical order to which Nicéphore belonged was suppressed in the early years of the Revolution, and he was forced to flee his home. He joined the military as an infantryman and served for several years until illness led to his discharge. While in convalescence, Nicéphore fell in love with Agnes Romero, one of the women who was taking care of him, and they were soon married. They ended up settling in Saint-Roch, a village near the larger city of Nice in southeast France, and had a son, Isidore, in 1795. Claude, who had become a sailor during the Revolution, joined them in Saint-Roch around the same time.

Apparently the quiet life did not suit the two restless brothers, who began a number of engineering projects. This work began in Saint-Roch, but in 1801, with the Revolution over, they decided it was safe to rejoin family back in Chalon-sur-Saône, where their research continued without pause. Their first noteworthy invention, completed in 1807, was incredibly ahead of its time: dubbed the *pyréolophore*, it was a primitive but functioning internal combustion engine, created some eighty years before the first automobiles went into production.

The family fortune had taken a tremendous blow over the course of the Revolution, so the work of the brothers was motivated not only by curiosity but by finances. They built their reputation by performing studies of textiles for the government, which earned them official praise. But their interest, and their destinies, were permanently steered toward photography after the introduction in France of another sort of "-ography": *lithography*. The term, which means "stone writing," refers to the method by which an image is affixed to a stone, allowing it to be reproduced countless times on paper.

In the original technique, an artist drew an image on a flat piece of stone using a soapy, greasy substance. The lithographer then treated the entire stone plate with a bath of weak acid and gum arabic; the gum was naturally drawn into the areas that were not covered with the grease, resulting in those areas becoming impervious to water. When ink was rolled onto the plate, it would stick only to those regions previously protected with the grease. The stone could then be placed on paper to print the embedded image, and it could be reinked and used over and over to make numerous copies.

Lithography revolutionized printing to an extent not seen since the original invention of the printing press. Now illustrations could be mass-produced for books and papers just as easily as typed words. The process was invented in Germany by the author and actor Alois Senefelder in 1796, but did not catch on in France until around 1813, when intellectuals and the wealthy began to experiment with it wholeheartedly. The Niépce brothers joined in on the new lithography craze, with Nicéphore's son, Isidore, providing the artwork that would be conveyed to the stone.

It seems that the departure of Isidore to join the army spurred the next stage of research. Missing an artist, Nicéphore and Claude began to wonder if they might be able to somehow directly capture a scene from nature. Using their chemistry knowledge, they began to experiment with materials to find one that was sensitive to light, with the goal of recording images produced by the camera obscura. Already by 1816, they had made progress in achieving crude images using silver compounds like the one that Johann Heinrich Schulze had baffled his friends with a century earlier. But producing an image turned out to be the easier task; making that image permanent was the real challenge. Without some sort of additional chemical treatment to stop the photosensitive process, any picture taken would eventually be degraded by further exposure to light. Such a "fixing" agent would elude the brothers for several more years.

Nicéphore ended up having to work on the process alone. Claude departed in 1816 for Paris, and then London, to find sponsors for their other masterwork, the pyréolophore combustion engine. Though the two brothers corresponded regularly about their respective progress, Nicéphore was the main force behind the imaging research. By 1820, he at last found an appropriate substance for recording the image, as well as one for fixing it. Bitumen of Judea, or Syrian asphalt—also used in lithography—was chemically modified by light exposure into a form that was resistant to being dissolved in petroleum. A film of this bitumen could be put on glass and exposed to light, becoming hard in spots where the light was brightest and remaining soft in the dark regions. Petroleum could then be used to wash away the soft regions. A crude but permanent image had been created. The Niépce brothers called the resulting image a heliograph, or "sun writing"—the word *photograph* did not appear until much later.

The oldest surviving heliograph dates from 1826 and, like most of Nicéphore's original images, shows a view from the window of his home. Even after the image is enhanced, we can see the limitations of the early process. The image is very grainy; the contrast between light and dark is extreme. The exposure time required was at least eight hours, possibly up to a full day, and was certainly far too long to capture anything moving—such as a falling cat.

The historian George Potonniée argues that the birth of photography, in the modern sense of the word, can be dated to 1822.[6] This was the year that Nicéphore perfected his process so that a permanent image could be chemically recorded via a camera obscura. Nicéphore continued to improve the technique in private over the next few years; he and Claude were deliberately vague in their letters to each other, lest any prying eyes intercept their secrets in transit.

But news of the invention accidentally slipped out in a fateful meeting in 1826. Nicéphore had asked one of his relatives, a Colonel Laurent Niépce, who was traveling through Paris, to pick up

The earliest surviving heliograph by Joseph-Nicéphore Niépce, "View from the Window at Le Gras," dated 1826 or 1827. The contrast has been enhanced dramatically by Helmut Gersheim.

a new camera obscura for him at the shop of the renowned Chevalier family. The colonel boasted to the Chevaliers about Nicéphore's work and even showed off one of the heliograph images. The amazed proprietors remembered that one of their customers, the artist Louis Daguerre, had been attempting to perfect a similar process for over a year. In short order, the Chevaliers told Daguerre about the conversation, and Daguerre wrote to Nicéphore asking for a scientific dialogue. The suspicious Nicéphore ignored the first letter entirely, but when a second letter arrived a full year later, in 1827, the two men started a cautious correspondence.

Louis Daguerre and the Niépce brothers came from much different backgrounds. The son of the clerk of a local bailiff, Daguerre was born in 1787 in a small village named Cormeilles-en-Parisis. His working-class status did not give him good educational opportunities, but he showed an aptitude for drawing at a young age and had an energy and determination that would help him overcome many obstacles in life. To foster Louis's artistic talent, his parents arranged for him to work with an architect in Orléans. His skill at that profession secured him a chance to move to Paris to work for a famed painter of opera scenery, where he went quickly from being an apprentice to being in charge of decorations. One of his great achievements, which would serve him well in his future camera work, was improving the lighting effects for the theater, in part by using clever optical illusions.

In 1822 the ambitious Daguerre joined forces with another painter, Charles Marie Bouton, to create a new theatrical experience that they named the Diorama. In modern times, the word *diorama* has come to refer to any three-dimensional model of a scene, but Daguerre and Bouton's Diorama was an attempt to convincingly simulate a large outdoor setting in an enclosed theater. Ordinary foreground objects were harmoniously arranged with skillfully painted structures mid-stage and with paintings in the background to create the illusion of a vast space. Daguerre used his experience with lighting to show transitions from day to night and to simulate atmospheric effects. The Daguerre-Bouton Diorama was an immediate success and made Daguerre very wealthy. In short order, he bought out his partners to run the enterprise on his own.

It was while working on the Diorama that Daguerre allegedly found the inspiration to study the photographic process. In the summer of 1823 he was working on a painting for the Diorama when he noticed the inverted image of a tree projected onto his painting. This image came via a small hole in a window shutter—an accidental camera obscura. The next day, when Daguerre returned to his work, he found that the image of the tree remained upon the canvas—an inadvertent photograph! His initial attempts to reproduce the effect ended in failure until he remembered that he had mixed iodine into the paints he had been using. From then on, he worked on recording images using iodine-based compounds.

The historian Potonniée thinks the story is a legend—it was recounted secondhand by a friend of Daguerre's—so there is ample reason to be suspicious. Daguerre may simply have been inspired to explore photography by his own use of the camera obscura in drawing landscapes. In any case, by 1827 he had become utterly obsessed with the possibility of photography, so much so that he was risking his marriage and his fortune. The French chemist Jean-Baptiste Dumas recollected the following.

> In 1827, when I was still young, I was told that someone wished to speak to me. It was Mme. Daguerre. She came to consult me on the subject of her husband's researches, which she feared would be a failure. She did not hide her anxiety for the future and asked me if there was any hope that her husband might realize his dream and, timidly, whether there was not some way to have him declared incapable of managing his affairs.[7]

Her worries would turn out to be unwarranted. In late 1827, Nicéphore and his wife, Agnes, took a trip to London to visit Nicéphore's brother, Claude, whose health had deteriorated significantly with the stress of seeking sponsorship for the pyréolophore. Claude died in early 1828, only days after Nicéphore returned to France. But along the way to London, Nicéphore and Agnes visited Daguerre in Paris. Claude's struggles and the heavy expense of funding heliograph research had put the Niépce family in financial peril, and an alliance with the wealthy Daguerre seemed to be a reasonable course of action. Nicéphore himself was sixty-three years old in 1828, and his energy and ability to do the hard experimental work on his own was failing. Though it took two more years of correspondence, on December 14, 1829, Daguerre and Niépce signed a collaborative agreement on photographic research. Nicéphore died only a few years later, in 1833, and his son, Isidore, took over what remained of the partnership.

It took several more years for Daguerre to develop a truly viable and successful photographic process. By 1835 he had found a reliable method to make positive images on a silver-coated metal plate, using iodine to help form the image and mercury to fix it. By 1839 he was ready to reveal the process to the world, spurred on financially by the costly destruction of his lucrative Diorama by fire. On June 14, 1839, the French government entered an agreement with Niépce and Daguerre to purchase the process in exchange for a lifetime pension for each of them, pensions that would extend to their widows. The images formed by this technique became known as daguerreotypes, and they were an instant international sensation.

In the short term, the Niépces' role in the development of photography was suppressed and ignored. In 1835, Daguerre more or less forced the financially strained Isidore Niépce to accept a new partnership that omitted the Niépce name from the business altogether; hence we have "Daguerreotypes" and not "Niépce-Daguerreotypes." When the French government and scientific community introduced the new process to the world, there was an active effort to minimize the Niépce contributions. The minister of the interior, having been taught the history of the process by the physicist (and friend of Daguerre) François Arago, had this to say in the 1839 announcement.

> M. Niépce, the father, invented a method of rendering these images permanent but, although he had solved this difficult problem, his invention, nevertheless, remained very imperfect. He obtained only a silhouette of the objects, and it required at least twelve hours to obtain any kind of design. Following an entirely different road, and putting aside the experience of M. Niépce, Daguerre has arrived at admirable results of which today we are the witnesses. . . . The method of M. Daguerre is his own; it belongs to him alone and it differs from that of his predecessor as well in its cause as in its effects.[8]

This story is unfair to Niépce, without whose initial work Daguerre may never have found his own process. The unfortunate Niépce family were pioneers in two world-changing technologies, photography and the internal combustion engine, but received little recognition or reward for either.

The Daguerreotype process required only minutes to capture an image, a tremendous improvement over the hours or days required to produce a Niépce heliograph. This was still, however, much too slow to properly record living creatures, much less those in rapid motion. One of Daguerre's surviving images from 1839, showing a view of the Boulevard du Temple in Paris, is eerie in that it was taken at midday but shows virtually empty streets—a city of ghosts. This is the earliest known candid photograph of a human being, however, for a man having his shoes shined in the left foreground, at the end of the row of trees, happened to stand still long enough to register an image.

The release of Daguerre's work to the public led to rapid improvements in the photographic process among enthusiastic scientists and entrepreneurs. Over the course of a few short years, the time to take a

Louis Daguerre, "Boulevard du Temple, Paris," 1839. Wikimedia Commons.

photograph went from minutes, to seconds, to a fraction of a second. Somewhere in this time frame, people started taking Daguerreotypes of beloved pets. Two images held by Harvard University's Houghton Library are candidates for the earliest surviving cat photograph; the one shown here was taken between 1840 and 1860. As with the Boulevard du Temple, only the stationary objects are well developed. The cat's head, moving while it eats, is blurry.

Other improvements in the speed, reliability, and popularity of photography came along quickly. In 1839, the same year that Daguerre released his process to the world, the British scientist Henry Fox Talbot revealed that he had been working on his own distinct photographic process since 1835. By 1841 he had improved the technique and named it the *calotype*; in Fox's method, images could be captured on paper within minutes in the form of negative images, allowing positive photographs to be easily duplicated many times from a single negative.[9] Portrait galleries opened, but even with

Candidate for the oldest cat photo. TypDAG2831, Houghton Library, Harvard University.

the increased speed of photography, subjects had to stand very still for an unblurred image to be taken. Postmortem photography, in which photographs were taken of deceased loved ones, also blossomed at this time; there was little threat of the subjects blinking on camera.

Everyone could see that the trend in photography was toward faster images, and could imagine the amazing possibilities that would become reality when images could be captured almost instantly. In 1871, for example, one John L. Gihon presented to the Pennsylvania Photographic Association an update on the state of the art of "Instantaneous Photography," which suggests that progress had already been made. "Some years ago, when working outdoors, endeavoring to obtain instantaneous effects, I used a simple contrivance that gave an exposure quick enough to enable me to catch the figure of an animal in rather rapid motion. A box was attached to the front of the camera. A board with a hole in it worked in a slide. This was operated upon by a trigger, and the dropping of the perforated board before the lens gave the exposure."[10]

Whereas early photography was limited by the speed of the chemical process on the film, now high-speed photography needed an additional technological advance: an automated shutter that could open and close within a tiny fraction of a second. Early photographic processes were so slow that the photographer could remove and replace the lens cap by hand to capture an image. To photograph objects in rapid motion, the chemical and mechanical processes of photography would both need improvement.

Photography was invented by a man of science, Niépce, and perfected and popularized by an artist, Daguerre. In a curious mirror-imaging of history, an artist would invent true high-speed photography, and a scientist would perfect it. The scientist in the latter case is the Frenchman Étienne-Jules Marey (1830–1904), who would revolutionize not only photography but physiology and produce the first photos of a falling cat. The artist in this case, who would go by numerous names during his lifetime, was born Edward James Muggeridge (1830–1904) but would achieve lasting fame as Eadweard Muybridge. He would definitively answer the question of how a horse really moves when it gallops.

Edward Muggeridge was born in Kingston upon Thames, England, and was later remembered by his family as an eccentric, energetic, and talented child.[11] Though only ten miles from the center of London, Kingston was a rather quiet locale, far enough from the city to be considered remote. Edward was not satisfied with the quiet life and by age twenty had moved to the United States to make a name for himself—literally, for he changed his surname to Muygridge in the States. He started in New York City as an agent for a publishing house, with the eastern and southern states as his territory, but by 1856 he had moved west to San Francisco and taken up business as a bookseller.

California was then, as now, seen as a place for people to reinvent themselves in a way that would bring them fame and fortune. It was booming in the 1850s, since the gold rush of 1848 had brought waves of fortune seekers, investors, and opportunists to the area. San Francisco became the nexus of activity, its population exploding from one thousand people in 1848 to twenty-five thousand in 1850. That same year, California jumped directly into statehood as part of a compromise between slave-owning states and free states; this change gave it more rights and recognition, which undoubtedly helped it grow in influence.

Given the excitement and opportunities in the new state, Muygridge must have concluded that bookselling was not compelling enough; in 1859 he sold the entire business to his brother and announced that he would be taking a trip across the United States and on to Europe, apparently to purchase more antiquarian books.

The trip went disastrously wrong in a manner that changed Muygridge's life and career. On July 2, 1860, he boarded an overland stagecoach heading to St. Louis, where he could catch a train to New York City. For the first few weeks, everything was routine. But the brakes of the coach failed after a stop in Texas, and the horses refused to slow, leading the driver to panic and attempt to halt the vehicle by steering it toward a tree. The resulting collision killed one man and injured everyone on board; Muygridge suffered a severe head injury. In the short term, he was plagued with double vision, confused thinking, and other cognitive problems; in the long term, the damage to his brain seems to have changed his personality, making him more erratic, compulsive—and dangerous.

After receiving a financial settlement from the stagecoach company, Muygridge returned to England to consult with top doctors about his ailments and was probably prescribed rest, relaxation, and outdoor activity. There is little record of what he did for the next seven years, but photography must have been part of his relaxation regimen. In 1867, Muygridge returned to San Francisco, now with the surname Muybridge and with a new calling as a professional photographer.

Muybridge specialized in landscape photography, and his base of operations in San Francisco placed him near one of the best subjects in the United States: Yosemite Valley. At that time, the valley was remote, access was difficult, and tourism was extremely rare. Interest in the valley's beauty was high among the public, however, and photographers flocked there to take advantage of the enthusiasm. Muybridge managed to distinguish himself in part by lugging his bulky equipment to dangerous viewpoints; he risked standing behind waterfalls and teetering over rocky crevasses to take unusual photographs. He also became noteworthy for capturing turbulent and cloudy skies in his vistas, something that was difficult to do with the technology of that era. An exposure that would develop a landscape would overdevelop the sky, and an exposure designed to capture the sky would leave the landscape looking dark. Muybridge overcame this technical difficulty at first by "faking it": taking separate photos of the land and the sky and superimposing them in a final print.

He soon, however, invented a device that would allow him to accurately capture earth and sky together at the same time, foreshadowing his later work in high-speed photography. The "sky shade" that he developed and reported on in 1869, using his photography pen name Helios, was essentially a piece of wood fitted into grooves that allowed it to rapidly fall into place in front of the camera lenses.

Falling from above, the shutter would block out the sky first, leaving more time for the landscape to be recorded on film.[12] Muybridge was already thinking about how photography could be improved through the use of novel mechanical shutters.

The year 1869 also saw the completion of a monumental and stunningly historic task in the United States: the building of the first transcontinental railroad connecting the east and west coasts. For years, the Union Pacific Railroad Company had been building westward, and the Central Pacific Railroad Company had been building eastward. The two rail lines met on May 10, 1869, at Promontory Summit, in the Utah Territory. The final ceremonial golden railroad spike was hammered into place by Leland Stanford (1824–1893), president of the Central Pacific Company and one of its original investors.

Stanford held many posts, identities, and interests during his lifetime. He worked in his early years as a lawyer in New York and Wisconsin, but after his law office burned in a fire, he moved west to California, following many others in the gold rush. Capitalizing on the needs of miners, he ran a successful general store, then invested the fortune he made into the railroad, further increasing his wealth and power. After an unsuccessful run for governor in 1859, he tried again, in 1861, and won, serving one term. In 1869, the same year he hammered in the golden spike, he founded a winery in Alameda County.

The stress and labor of his railroad work—as well as the corrupt and cruel business practices he engaged in to get it done—evidently took a toll on Stanford's health. His doctor recommended that he indulge in leisure travel for relaxation, but he was unable or unwilling to leave California. As a compromise, he took up the purchase, breeding, and racing of horses, and in 1870 he purchased Occident, a champion trotting horse—a significant move, it turned out, for the investigation of falling cats.

Trotters typically pulled a two-wheeled cart behind them and raced using a gait slower than a gallop but still too fast for viewers to see the horses' legs in motion. In the world of racing, Stanford waded into a debate that had been raging among the wealthy owners: Does a trotting horse ever have all four legs off the ground at the same time? With the many improvements in photography, it must have seemed natural to Stanford to see if photographs could answer the question once and for all, so in 1872 he hired Muybridge to solve the mystery, providing funding, facilities, and horses for the artist to accomplish the task.

It is not known exactly why Stanford chose Muybridge. It has been suggested that it may have been Muybridge's advertisement of his capabilities: "HELIOS is prepared to photograph private residences, *animals*, or views in the city or any part of the coast."[13] Once the two men met, however, it seems likely that Stanford saw a kindred spirit in Muybridge: a bold and imaginative risk taker with large ambitions.

The choice of Muybridge was a wise one. By early 1873, Muybridge had successfully taken a photograph of Occident trotting with all four feet off the ground. To accomplish this, he created his own high-speed shutter to shorten the length of exposure. To overcome the slow speed of the wet-plate chemical process he was using, he blanketed the area to be photographed with white sheets, providing lots of reflected light. Occident appeared only as a silhouette in the photographs, but this silhouette was enough to answer the question of horse trotting.

Muybridge's accomplishment did not leave a lasting impression on the photographic world, at least at that time. The image was poor, and some even considered it a fake. Though Muybridge had taken multiple photographs of the horse, it was not possible to arrange them in a sequence to show the actual dynamic motion of a horse, and they were thus only useful in answering the original debate question. With the problem solved to his satisfaction, Muybridge continued with his more conventional photographic studies in Yosemite and took a photography commission to document the Modoc War in 1873, in which Native Americans fought against their unjust removal from their hereditary home.

Muybridge had made a very good name for himself in photography. It is the darkest of ironies that a photograph would also lead him into madness and murder. On October 17, 1874, Muybridge made a fateful and accidental discovery in his own home. He came across an image of his son Florado, born seven months earlier to his wife, Flora. The photo itself was ordinary; on the back of it, however, he found written, in his wife's handwriting, "Little Harry."

Muybridge immediately recognized the "Harry" in this case as Harry Larkyns, a handsome scoundrel and con man who had made the acquaintance of the Muybridge family a year earlier. As was common societal practice in San Francisco, Larkyns became a social companion to Flora at public events, escorting her to the theater and various parties; Muybridge himself disdained such gatherings and was happier spending his evenings on photographic work. But the implication of the photograph of Florado was obvious: Flora and Harry had a more than friendly relationship, and Muybridge's son might not be his own.

The day of the discovery, Muybridge wandered about town in visible torment, asking friends to keep his affairs in order. That afternoon he caught a ferry out of San Francisco to Vallejo, where he took a train to Calistoga—the resort town where Larkyns was staying. There, after making inquiries, he took a carriage to the home where Larkyns was a guest, entered a gathering in the parlor, and shot Larkyns dead.

The trial that followed was a regional sensation. Muybridge's lawyer put forth an insanity defense and, remarkably, on February 5, 1875, Muybridge was found not guilty. It seems likely that the legal defense resonated less with the jury than did the social mores of the time. A brutal premeditated murder might have seemed a reasonable reaction to cuckoldry. Muybridge was lucky, but he passed none of that good fortune along to his wife. Flora, denied both financial support and a divorce from her bitter husband, fell into ill health and died in July 1875; her son was sent to an orphanage not long after. By the time of both tragedies, Muybridge was in Guatemala on a photographic expedition.

By 1877 he had returned to San Francisco. Among numerous other photographic projects, he took up collaboration with Stanford again to study animals in motion. Stanford had expanded his horse racing and breeding operations at his various properties. The new goal of the research was not to settle gentlemen's bets but to study the motion of horses with an eye toward improving their speed and efficiency. In 1877, with advances again in camera shutters and chemical processes, Muybridge produced better instantaneous images of Occident at a trot; they were still poor, but they showed that the technique had potential.

The final breakthrough came with the help of Stanford's considerable railroad resources. Stanford directed his railroad engineers to help Muybridge with any technical support needed, and near the end of 1877 the engineers had developed an electrical shutter system that could be triggered when a subject such as a horse broke a wire as it passed. With the new shutter, Muybridge was able to set up a row of

Eadweard Muybridge, "The Horse in Motion," 1878. Library of Congress Prints and Photographs Division, Washington, D.C.

cameras along the racetrack, up to twenty-four in the end, and get a consistent time series of a horse in a trot or a gallop.

Today the images of Stanford's horse Sallie Gardner in a full gallop are remembered best, though at the time it was the trotting images that seemed to resonate most with the public. Muybridge's accomplishment gave him near-instant worldwide fame; almost nobody questioned the importance of his work both for science and for art. In an 1878 issue of *Scientific American*, the achievement is described in almost giddy terms.

> The most careless observer of these figures will not fail to notice that the conventional figure of a trotting horse in motion does not appear in any of them, nor anything like it. Before these pictures were taken no artist would have dared to draw a horse as a horse really is when in motion, even if

> it had been possible for the unaided eye to detect his real attitude. At first sight an artist will say of many of the positions that there is absolutely no "motion" at all in them; yet after a little study the conventional idea gives way to truth, and every posture becomes instinct with a greater motive than the conventional figure of a trotting horse could possibly show. Mr. Muybridge's ingenious and successful efforts to catch and fix the fleeting attitudes of moving animals thus not only make a notable addition to our stock of positive knowledge, but must also effect a radical change in the art of depicting horses in motion. And everyone interested in the physiology of animal action, not less than artists and horse-fanciers, will find the photographs of Mr. Muybridge indispensable.[14]

The question of the galloping horse had been solved by Muybridge, throwing artists into a fierce debate over whether this knowledge was a boon or a burden to them. As Lankester noted in 1913,

> How, then, we may now ask, ought an artist to represent a galloping horse? Some critics say that he ought not to represent anything in such rapid action at all. But, putting that opinion aside, it is an interesting question as to what a painter should depict on his canvas in order to convey to others who look at it the state of mind, of impression, of feeling, emotion, judgment, which a live, galloping horse produces in him.
>
> . . .
>
> But there is, further, in all "seeing" before even a mental result of *attention* to the retinal pictures is, as it were, "passed,"

Giacomo Balla, "Dynamism of a Dog on a Leash," 1912. Albright-Knox Art Gallery, New York.

admitted and registered as a "thing seen," the further operation of rapid criticism or *judgment*, brief though it may be. We are always unconsciously forming lightning-like judgments by the use of our eyes, rejecting the improbable, and (as we consider) preposterous, and accepting and therefore "seeing" what our judgment approves even when it is not there! We accept as "a thing seen" a wheel buzzing round with something like fifty spokes—but we cannot accept a horse with eight or sixteen legs! The four-leggedness of a horse is too dominant a prejudice for us to accept a horse with several indistinct blurred legs as representing what we see when the horse gallops.[15]

Indeed, it can look rather peculiar to represent an animal's motion in still form this way. In 1912 the artist Giacomo Balla, inspired by the motion studies of Muybridge and others, painted "Dynamism of a Dog on a Leash," in which a little dachshund looks as though it had just been startled by a ghost and is cartoonishly trying to run away on an icy surface. Artists would have to come to grips with the fact that "realism" often doesn't look real, at least where paintings of objects in rapid motion are concerned.

At the time of Muybridge's photographs, reports suggest that some artists took the revelations much harder than others. The French painter Jean-Louis Ernest Meissonier had long prided himself on accurate depictions of animals, especially horses, in motion and had won a minor battle in the art world in getting his depiction of a trotting horse accepted as the standard pose in painting. In 1879, Leland Stanford visited Meissonier, with the goal of getting the famed artist to paint his portrait. Meissonier was reluctant until Stanford brought out the photographs of the horse in motion.

> Meissonier's eyes were filled with wonder and astonishment. "How!" he said, "all these years my eyes have deceived me!" "The machine cannot lie," answered Governor Stanford. The artist would not allow himself to be convinced, and, rushing to the other room, brought forth a miniature horse and rider made of wax and by his own hands. Nothing could be more perfect, more beautiful, than this statuette.
>
> . . .
>
> It was almost pitiful to see the old man sorrowfully relinquish his convictions of so many years, and tears filled his eyes as he exclaimed that he was too old to unlearn and begin anew.[16]

This quotation, presented from a source identified only as a "Paris letter writer," may have been overstating Meissonier's reaction. Since it is unclear that anyone would have been in the room other than Stanford and Meissonier, it may well have been a story leaked by Stanford himself to promote his own accomplishments. The photographs served their purpose with Meissonier, however, who agreed to do Stanford's portrait in exchange for more photographs of animals in motion.

So the most immediate impact of Muybridge's photographs seems to have been felt in the art world. The 1878 *Scientific American* article, however, was prescient in noting that "everyone interested in the physiology of animal action . . . will find the photographs of Mr. Muybridge indispensable." Only a couple of months after it appeared, a letter appeared in the December 28, 1878, issue of the French journal *La Nature*, proving the point.

> Dear friend,
>
> I am in admiration of the instantaneous photographs of M. Muybridge, which you have published in your last number of *La Nature*. Could you put me in touch with the author? I should like to ask him to assist in the solution of certain problems of physiology very difficult to solve by other methods.[17]

This letter was from the French physiologist Étienne-Jules Marey, who would transform high-speed photography into a revolutionary scientific tool for studying animal motion—and would use that tool to baffle physicists with the problem of the falling cat.

4
Cats on Film

Eadweard Muybridge, the artist, had produced the first series of instantaneous photographs and used them to view the motion of living creatures. Étienne-Jules Marey, the scientist, would develop this technique into the most rigorous of scientific tools for the study of the motion of animals, people, and objects. In the process, he would lay the foundations for the motion picture industry. While building his incredible body of scientific motion studies, he would—almost as an afterthought—illustrate the motion of a falling cat and stun the worldwide physics community.

Marey was born in 1830—the same year as Muybridge—in Beaune, France, the only child of Marie-Joséphine and Claude Marey. The family was reasonably wealthy, for Claude worked as a wine steward in Beaune, a major wine-producing town. In a story repeated several times already in this book, the young Étienne-Jules showed intelligence and a mechanical aptitude at a young age: he had "brains in his fingertips."[1] When he went to a local college at age eighteen, he not only excelled in his coursework and won many prizes but made many friends by using his mechanical skills to make toys for his classmates.

With his passion for, and skill with, machines, Marey was interested in entering school to study as an engineer; however, and again we see a story repeat itself, his father wanted him to become a doctor.

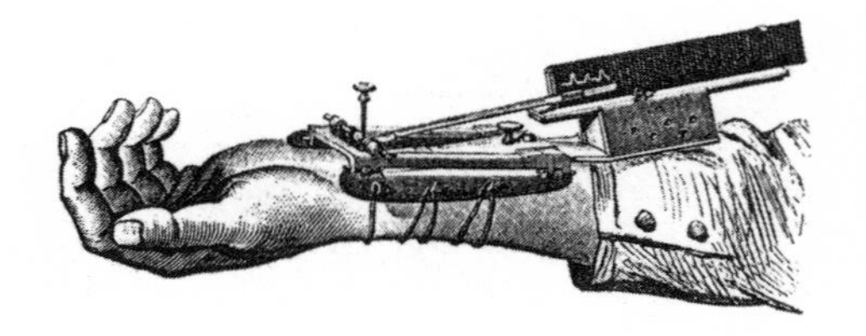

Étienne-Jules Marey's sphygmograph. From Marey, La méthode graphique, *p. 560.*

So, in 1849, Étienne-Jules entered medical school in Paris. In spite of its being his second choice of career, he mastered his medical studies and stood out as an innovative and imaginative thinker.

A turning point for Marey seems to have been an opportunity that he received in 1854 to work for a time in the laboratory of Dr. Martin Magron, a physiologist. Physiology may be defined as the study of how living organisms and their interconnected parts operate; it includes investigations in how the joints, muscles, and organs of a living creature perform their functions. This was a natural fit for the mechanically inclined Marey, who ended up specializing in studies of blood circulation. He defended his thesis, "The Circulation of Blood in Normal and Pathological States," in 1859 and developed his first diagnostic medical device soon after: the sphygmograph, which could directly measure, and plot on a piece of paper, the pulse of a human being at the wrist.

The pressure from the human pulse would move a stylus up and down, where it would trace out a pattern on a piece of soot-covered paper that was moved by a clockwork mechanism. Others had measured the pulses of animals before, but Marey was the first to design a device that could measure the pulse without requiring an invasive probe to be inserted into the subject. The sphygmograph was immediately recognized as an important medical tool and was put into widespread use. In a rather macabre twist, Marey's fortunes were further bolstered by the misfortune of another, as recounted here.

> M. Brouardel once told me that Napoleon III, having heard of the sphygmograph, sent for Marey, who was asked to make some experiments. He made some, and among the [pulse] lines he took from those present, he noticed a line that clearly indicated a marked aortic insufficiency. A few days later, the subject who had provided this pathological trace was found dead in bed, having succumbed to one of those syncopal attacks so frequent in heart disease that the sphygmograph had revealed to Marey.[2]

Marey's medical career had some significant setbacks around the same time. After defending his thesis, he passed an accreditation exam that qualified him to work as a doctor but failed another that would have allowed him to teach medicine. He consequently opened a medical practice in Paris, but it failed within the year. With no other options available, he set himself up as a private researcher of physiology; the work was funded by royalties from the sphygmograph invention and by the private tutoring of students. His combined living and working area housed a wondrous menagerie of creatures waiting their turn to be studied in motion. A colleague who visited him in 1864 described the space.

> Not only a laboratory but also a menagerie, the place was unforgettable. My first sight of it made such an impression that the memory remains among those that, upon evocation, will forever reveal themselves like the still-damp ink of some freshly printed etching.

> In impeccable order, the sort in which all urgent work finds itself discomfited, among scientific apparatus and instruments of all kinds, classical or invented just yesterday—for the new science, new tools—were cages, aquariums, and beings to populate them: pigeons, buzzards, fish, saurians, ophidians, amphibians. The pigeons cooed; the buzzards breathed not a word, perhaps for fear of rebuke of their qualities or reputations as buzzards. A frog, escaped from its jar in an exceptional offense, leaped dazed before the visitor to escape the sole's caress. A tortoise plodded, weighty with gravity, with a stubborn constancy more characteristic than vain eagerness, forced into a meandering route by various impedimenta, indefatigable in pursuit of its task as though driven by obsession, enjoying the security of both a clear conscience and its shell.[3]

Marey's work included lectures and publications on the cardiovascular system, and his efforts were recognized by the French scientific community, bringing him election to academies, assistantships, and a professorship in short order. By 1870 he was wealthy enough to purchase a house outside Naples, Italy, so he could comfortably continue his research during the winter months. As the years passed, he would receive increasingly more recognition for his work and more resources to carry it out.

But what was this work? Marey viewed motion as the key to understanding nature, for everything—atoms, planets, horses, people—is in motion and subject to the same physical laws. Unlike many of his colleagues in physiology, Marey did not hold that living creatures possessed some sort of special "vital force" that was not subject to the laws of nature. Ever the mechanical engineer, Marey felt that living creatures could be understood by the same techniques used in the physical sciences. As he would cautiously write, "Without doubt, however, there are numerical relations between the phenomena of life; and we shall arrive at the discovery of them more or less speedily, according to the exactitude of the methods of investigation to which we have recourse."[4]

To find those numerical relations, Marey developed instruments along the lines of his sphygmograph of 1860, which could create a graph of a person's blood pressure over time. Marey wanted to find precise ways to graph and measure all motions of animals, both external movement and internal activity. He was exceptionally successful at these efforts. Most relevant to our interests are his studies of the ways animals move themselves. For humans and horses, he set up a series of pneumatic tubes that went from the feet of a running creature to a handheld device to which it transmitted the air pressure of the foot's impact. The device recorded when, and for how long, each foot remained on the ground.

Marey's data, at the bottom of his illustration, included here, shows the impacts of a trotting horse's hooves as time advances to the right. The horizontal gaps between the impacts show that a trotting horse does, in fact, have all four feet off the ground at the same time. Marey's publication of this result, in the French edition of *Animal Mechanism*, came out in 1872; Muybridge's first photographs of Occident trotting came out in 1873. Marey had beaten Muybridge in answering the question of the horse's trot! His conclusions, however, do not seem to have convinced non-scientists, perhaps because Marey's graphical method did not have the same visual appeal as Muybridge's photographs.

Marey differed in his methods from many of his colleagues in that he was strongly against the practice of vivisection to understand

A trotting horse equipped with sensors. The chart at the bottom shows the impacts of the hooves over time. From Marey, Animal Mechanism, *p. 8.*

animals' anatomy and organ functions. Where his contemporaries had no difficulty in slicing open living creatures to study their workings, Marey firmly believed that such techniques gave incorrect scientific results. The action of restraining and operating on an animal changed its natural operation, he thought, making any conclusions drawn from such studies suspect. Marey endeavored to record the processes of life in as unobtrusive a manner as possible.

He took this philosophy to studying creatures in flight. To investigate the motion of insect wings, he held an insect in place and allowed it to flap its wings freely against a rotating cylinder covered in soot. The wingtip removed the soot, leaving behind a trace showing how the wings had moved in time. Birds could not be held and studied in this way, so Marey developed a pressure-sensitive apparatus, similar to that used on horses, to measure the up-and-down swings of a bird in flight. The bird could fly freely, but would be held in a harness with tubes hanging down from it. Inescapably, the scientific apparatus was interfering with and likely changing the bird's natural motion. What Marey needed was a way to record the motion of the bird without touching it at all.

We can almost imagine Marey falling out of his chair when he first read the 1878 account of Muybridge's horse photographs. In photography Marey saw the potential answer to all of his problems. Four days after he received the issue of *La Nature* with Muybridge's images, Marey excitedly wrote to the magazine, as we saw. Muybridge was very willing to help pursue further motion studies.

> I have read with keen interest the letter . . . concerning the photographs representing the movements of the horse, which you have done me the honor to reproduce in your estimable journal; the complimentary remarks you have made there have caused me great pleasure. Please be so kind as to convey the assurance of my esteem to Professor Marey and to tell him that the reading of his famous work on the animal mechanism has inspired Governor Stanford with the first idea of the possibility of solving the problem of locomotion with the help of photography. Mr. Stanford consulted me on this subject, and at his request I resolved to assist him in his task. He charged me to pursue a series of more complete experiments.
>
> . . .
>
> In the beginning, we did not study the birds in their flight; but, Professor Marey having suggested the idea, we will also direct our experiments on this side.[5]

Muybridge sent along a new set of photographs of animals in motion for Marey to ponder. After this first promising correspondence, there is no evidence of further contact for over a year. Each of the men had local projects to consider, projects that were laying the foundations for the twentieth century. Muybridge, now at the height of his fame, was not only expanding his own research but traveling as a lecturer. He had developed a device, the zoopraxiscope, that could animate and project his motion studies for enthusiastic crowds, a precursor to the modern motion picture. Marey, in Paris, was busy in negotiations with the government for a new research complex. At the same time, he was investigating methods of flight in the manner most familiar to him: by building mechanical models. With a colleague he built not only bird models that could flap their wings but also fixed-wing airplane models that used compressed air to power propellers.

Finally, in 1881, Muybridge was able to make a short European tour, thanks to the insistence of the artist Meissonier and his negotiations with Muybridge's patron, Stanford. Marey gave the San Francisco photographer a grand welcome.

> M. Marey, Professor of the College of France, yesterday invited to his new house in the Trocadero, Boulevard Delessert, some foreign and French savants, together with his intimate friend, our director, M. Vilbort. The attraction for the evening consisted of the curious experiments of Mr. Muybridge, an American, in photographing the movements of animated beings.
>
> . . .
>
> Mr. Muybridge, an American savant, gives us the first experience of something that should be accorded to the whole

> Parisian public. He projects upon a white curtain photographs showing horses and other animals going at their most rapid gaits. But that is not all. His photography having taken "on the wing" each movement of which each gait is composed shows us the animal in the positions that our eye, taking in only the general ensemble, would not otherwise observe. M. Marey lent his cooperation to Mr. Muybridge and made witty remarks on each tableau.
>
> . . .
>
> The sitting was prolonged until late, but we regretted that the time had come, when it did, to bid adieu to M. Marey and Madame Vilbort, who did the honors of the evening so charmingly.
>
> Finally, let us address an inquiry to Messieurs Marey and Muybridge. Would it not be practicable by the zoetrope process to give a little additional speed to the Paris carriage horses? We should not require elegance of movement, but hurried journalists would be under lasting obligations to the inventor of some such contrivance.[6]

Pleas of journalists aside, perhaps the most striking detail is the close association attributed to Marey and Madame Vilbort at the party. This was not coincidental: in a curiously distorted mirror image of Muybridge's life, Marey was having an affair with the wife of the newspaper director Vilbort, apparently with the latter's acceptance. Marey even fathered a child, Francesca, by Madame Vilbort. Though he did not publicly acknowledge her as his daughter, he did bring her into Parisian society as his niece. The home that Marey purchased in Naples was chosen in large part because Madame Vilbort was staying there to recover from illness.

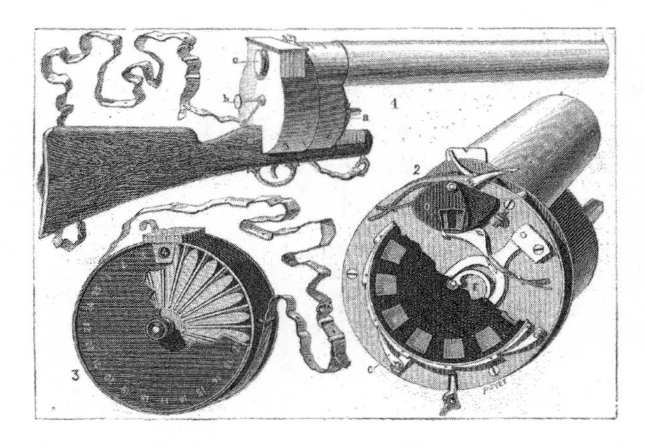

The fusil photographique. *From Marey, "Le fusil photographique." Reprinted courtesy of Cnum–Conservatoire numérique des Arts et Métiers.*

In his visit to Marey, Muybridge brought with him the requested photos of birds in flight. They proved disappointing and inadequate for Marey's purposes, for the images were not well formed, nor could the viewer distinguish the individual motions of the wing. Marey resolved to continue his investigations with his own custom equipment, and by early 1882 he had developed a *fusil photographique*—"photographic gun"—that could shoot a series of images on a filmstrip.

Marey might have been one of the first people to perform a "photo safari," shooting an animal with a camera rather than a gun. Other camera afficionados had made camera guns before, but theirs took a single image whereas Marey's could take a sequence. Marey's bang-less shooting drew the attention of locals near his Italian home in Posillipo. They reportedly referred to him as "lo scemo di Posillipo"—"the silly from Posillipo."[7]

Sequence of a running man. This drawing was reproduced from the original photograph; photographic reproduction in print was still not available. From Marey, "La photographie du mouvement." Reprinted courtesy of Cnum–Conservatoire numérique des Arts et Métiers.

The photographic gun was a great convenience and an advancement in motion photography, but it was an unstable platform for filming and produced separate photos around a disk that had to be cut and arranged to view properly. By mid-1882, Marey had developed another camera that could be fixed in place to capture all stages of an animal's motion on a single photographic film. This camera used a single unmoving photographic plate that was exposed to the target at successive instants by slots cut in a rotating disk. Marey presented his first results with the new camera in July 1882, depicting in a single image all aspects of a running man's motion. Marey would soon dub his new photographic method, applied to physiology, as *chronophotography*.

Such research, and the facilities to house it, would have cost a lot of money, more than Marey could have supported on his own. Fortunately, in 1880 he made the acquaintance of Georges Demeny, a patriotic young man who sought to improve the physical fitness of his compatriots. The French had suffered a serious defeat at the hands of the Prussians in the Franco-Prussian War (1870–1871). The birthrate of the country slumped after the war, and there was a real fear that the people of France were declining physically and morally. Physical education and fitness were considered to be critically important to avoid future defeats, and the military invested heavily in them. Marey had already been approached by the Ministry of War to apply motion studies to improving the health and fitness of soldiers; when Demeny asked to help with such work, Marey saw both a talented collaborator and someone who could handle the mundane practical aspects of physiology, leaving him free to pursue pure science.

After much bureaucratic wrangling, Marey established a permanent Station Physiologique on the western edge of Paris at the end of 1882. Now he had the labor, land, and resources to pursue almost any challenge that intrigued him. One limitation of Marey's newest chronophotography camera was that slow-moving objects produced images that overlapped. For instance, a man hardly travels any distance when he walks a single step: analyzing his gait via chronophotography resulted in an indistinguishable smear of images. Ingeniously, Marey realized that he did not need, or even want, to see the entire person in any image. By dressing a person in dark clothes and drawing simple white lines along the arms and legs, the motion could be visualized without confusion. Marey's strategy is similar to that used today in digital motion capture, in which actors have strategic dots placed on their bodies; these dots can be used in post-processing to determine the spatial position and orientation of the actors.

In 1884 the ever-busy Marey took a short break from photography to investigate a cholera outbreak that was rampaging across

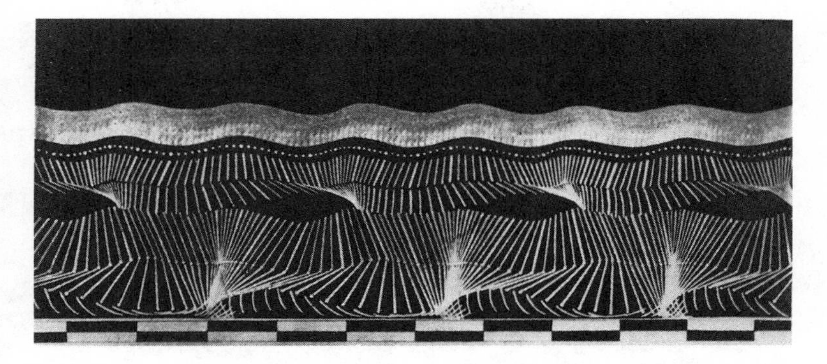

Sequence of a walking soldier by Marey, 1883. Courtesy of Iconotheque de la Cinematheque.

France. Marey and the microbiologist Louis Pasteur led a committee that worked to trace the origins of the epidemic and stop it. Using statistics and careful mapping, Marey and Pasteur convinced other members of the committee that contaminated water was how the disease spread.

When Marey returned to the Station Physiologique, which he had left in Demeny's care, he resumed his studies of animals. His "stick-figure" solution to capturing slow-moving objects was still not sufficient for many situations of interest. Photographing from in front of or behind a moving animal could also provide important information, but the animal would always appear in the same location in the camera's field of view, and the result would be a blur. Furthermore, there were stationary activities that Marey wanted to study, such as a boxer throwing a punch, a musician playing a violin, and a soldier lifting weights. In all these cases, the subject would not be walking or running, so the existing camera would not be effective. Since the subject would not move, the obvious solution was to move the film within the camera; however, Marey found that the glass film

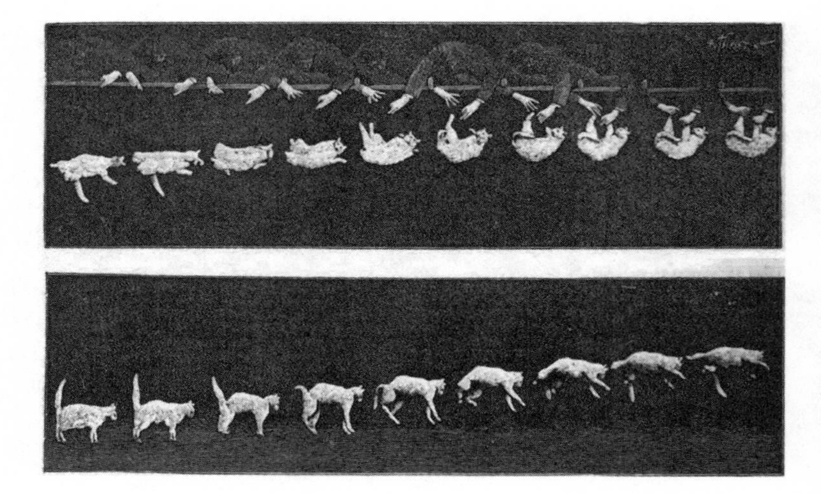

Side view of the gardener's falling cat, 1894. The images are read right to left, top to bottom. From "Falling Cat," a short film by Marey. Wikimedia Commons.

plates still in use could not be easily and reliably moved during a photographic sequence, and the images that resulted did not meet his high quality standards.

This time, the solution came from an external source. In late 1888 paper photographic films arrived in France from the United States, where they had been developed by George Eastman in Rochester, New York. Now Marey could remodel his camera to unroll a spool of film in front of the camera lens, resulting in a time sequence of images across the film strip—a design very close to that of a motion picture camera. With this advance, there was almost nothing that could not be recorded in motion. By 1892, Marey was filming the actions of any sort of animal he could get his hands on, from goats to dogs to insects to ducks to horses.

The falling cat seems almost to have been an afterthought in the research. The subject was not of military importance, like the motion studies of soldiers, nor of economic importance, like the studies of moving horses (thanks to everyone ignoring Niépce's internal combustion engine, automobiles were still a few years away from popular use). The gardener's cat at the Station Physiologique was drafted in the name of science and dropped on camera in 1894. The sequence was presented by Marey at the French Academy of Sciences on October 22, 1894.[8]

If Marey considered the photographs of the falling cat to be a simple demonstration of his new photographic capabilities, the reaction was quite disproportionate. His images were published all over the world and outraged physicists at the French Academy meeting when he presented them, as was later reported.

> Why does a cat always fall on its feet? This is a question which has recently absorbed the earnest attention of the French Academy of Sciences. The problem is clearly a difficult one, for that learned body of savants has so far failed to offer a final solution.
>
> . . .
>
> When M. Marey laid the results of his investigations before the Academy of Sciences, a lively discussion resulted. The difficulty was to explain how the cat could turn itself round without a fulcrum [point of leverage] to assist it in the operation. One member declared that M. Marey had presented them with a scientific paradox in direct contradiction with the most elementary mechanical principles.[9]

5

Going Round and Round

Sometime between Parent's work in 1700 and Marey's photographs in 1894, the physics of the falling cat had gone from being a solved problem in elementary physics texts to being a "scientific paradox." But in those two hundred years, physics had changed significantly, and it was now defined as much by what *is not* possible as by what *is* possible.

The key to these changes was the discovery and broad recognition of *conservation laws*, which indicate that some physical quantities cannot change in an isolated system. The most famous of these is the *law of conservation of energy*, which states that energy cannot be created or destroyed, merely changed from one form to another. Forms of energy include the energy of moving objects (kinetic energy), the energy stored in a gravitational field (gravitational potential energy), heat energy (the energy of a large number of particles moving randomly, as in a gas), chemical energy (energy stored in chemical bonds in molecules and atoms), and electromagnetic energy (the energy contained in light, ultraviolet radiation, infrared radiation, radio waves, and X-rays). In Einstein's special theory of relativity, it is further recognized that mass is itself a form of energy.

An example of the conservation of energy can be found in the operation of an automobile. A car is set into motion by converting chemical energy (from gasoline) into kinetic energy, with some amount of the chemical energy unavoidably converted into heat energy. In going uphill the car slows down as the kinetic energy is converted into gravitational potential energy, and the opposite happens when it goes downhill. When the driver steps on the brakes, the car's kinetic energy is converted into heat because of the friction between the wheels and the brake pads.

Hints of the existence of some sort of energy conservation principle stretch back to the time of the ancient Greeks, though practical formulations of the idea did not begin to appear until around the time of Isaac Newton.[1] Newton's rival, Gottfried Leibniz, first attempted to quantify the energy of moving objects; he called kinetic energy the *vis viva*, or "living force," of the system. But his vis viva seemed to be conserved only for astronomical bodies, like planets in motion, and not for bodies in motion on the Earth; scientists had not yet recognized heat as a form of motion.

The modern law of energy conservation appeared in the mid-1800s through the work of a pair of unlikely investigators: the German physician Julius von Mayer (1814–1878) and the British brewer James Prescott Joule (1818–1889). Mayer came by his insight while working as a ship's doctor on a Dutch vessel that traveled to the East Indies in 1840. While practicing bloodletting on some sick sailors, Mayer noticed that the blood emerging from the patients' veins was much redder than he would have expected; it looked more like the heavily oxygenated blood from arteries. It occurred to him that, in the high-temperature tropics, the human body did not need to consume as much oxygen from the blood to maintain its normal body temperature; thus the venous blood was redder, and more oxygen-laden, than it would be in cooler climates. Mayer realized that there was a balance of some sort of "energy" between the human body and the environment around it, and he insightfully recognized that this principle might apply to all sorts of physical processes. Local sailors bolstered his hypothesis by noting that the ocean temperature after a storm was higher than before it; the motion of the water caused by the winds was being converted into heat.

Joule came by his revelation while working to optimize his brewery's operations; his original goal was to compare different types of engines. He had been using steam engines in his brewery but wanted to determine whether newly developed electric motors might be more efficient and therefore more cost-effective. Though his investigations started as a purely practical problem, Joule became fascinated with the question of how energy can be converted from one form to another. He determined the *mechanical equivalent of heat*—how much mechanical work would be required to produce a given amount of heat—and presented his results to the British Association for the Advancement of Science in 1843, where he was met with stony silence. Mayer, who had published his results in 1841 and 1842, had found even more resistance to his ideas. Within a few years, however, physicists had demonstrated convincingly the relationship between, and interchangeability of, various forms of energy. From 1847 on, the conservation of energy was largely accepted.

One of the theoretical consequences of the law of conservation of energy was the death, in the scientific world at least, of the idea of "perpetual motion machines," engines that can be set into motion and run forever. Energy conservation indicates not only that there is a finite "well" of energy for any isolated machine, but that this machine will inexorably convert its energy into unusable heat. This conclusion did not keep one 1897 writer from, at least tongue in cheek, suggesting that cats can be used to create perpetual motion.

> A new industry will be started at Freeport, Illinois, says an exchange, on a quarter section of land. An enterprising farmer

> will establish a thousand black cats, and five thousand rats on which to feed the cats, estimating that the cats will increase fifteen thousand in two years, their skins being worth a dollar each. The rats will multiply five times as fast as the cats and will be used to feed the cats, while the skinned cats will furnish food for the rats. Thus has perpetual motion been discovered at last.—*Lippincott's Magazine*
>
> Discovered nothin! Nature has had this same patent double-back action rat and cat perpetual motion going on ever since Noah was a sailor.[2]

We can readily see how this scheme would fail, even without thinking too abstractly about conservation of energy. Even if every part of a rat is fed to the cats, not every part of the cat is being fed to the rats. There is inevitably a loss of mass in this system, and the farmer would do well to learn a little physics.

Though conservation of energy took quite some time to be recognized, another conservation law—the conservation of momentum—can be found directly in Isaac Newton's famous laws of motion. We may summarize these laws, which appeared for the first time in somewhat different form in Newton's *Principia*, as follows.

1. An object remains at rest or in constant motion until subject to an external force: the law of inertia.
2. The sum of external forces on an object is equal to the mass of the object times its acceleration: (force) = (mass) times (acceleration).
3. When one object exerts a force on a second object, the second object exerts an equal and opposite force back on the first: For every action, there is an equal and opposite reaction.

The *momentum* of an object, defined as "(mass) times (velocity)," may be roughly described as how much oomph an object has. If two objects have the same velocity and different masses, the object with the larger mass has more momentum; if two objects have the same mass and different velocities, the object with the larger velocity has more momentum. When a car and a truck collide on the road, the truck usually crushes the car because it has larger mass and typically more momentum.

Newton's laws suggest, in an indirect way, that momentum is conserved in any isolated physical system. Newton's first law tells us that the velocity of an object doesn't change unless it experiences an outside force; therefore the momentum of an object in isolation will not spontaneously change. Since the acceleration of an object is the rate of change of its velocity, Newton's second law tells us that a force represents a change in momentum. Newton's third law tells us that if the momentum of one object is changed, another object must have its momentum changed in exactly the opposite way, leaving the total momentum the same.

Billiard balls are often used to demonstrate conservation of momentum. If the cue ball is hit directly at the eight ball, the cue ball will stop and the eight ball will carry on in the same direction and with the same speed as the cue ball; the cue ball's momentum has been completely transferred to the eight ball.

Momentum is often called *linear momentum*, to distinguish it from a third quantity that also satisfies a conservation law: the *momentum of rotation*, or "angular momentum." Momentum of rotation represents, loosely, how much oomph an object in rotational motion has, including both objects that are spinning, like a bicycle wheel, and objects in orbit, such as the Earth moving around the Sun. For a point-like mass, the angular momentum satisfies the formula

Three wheels of equal mass:

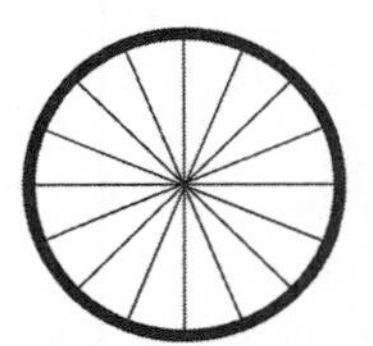

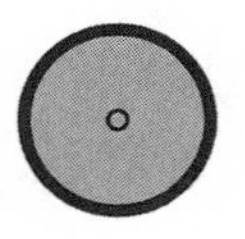

Highest moment of inertia: mass distributed around edge of large wheel

Smaller moment of inertia: mass distributed around edge of smaller wheel

Smallest moment of inertia: mass distributed through entire area of wheel

Three wheels with different moments of inertia, for comparison. My drawing.

"angular momentum = (radius) times (mass) times (velocity)," where *radius* is the radial distance at which the object is orbiting the point-like mass. From this, it can be seen that if two objects have the same mass and velocity but orbit at different distances, the one making the larger circle will have the larger angular momentum.

For a non-point-like spinning object, the implication of the formula is that angular momentum depends not only on the mass of an object but also on the distribution of the mass in the object. Both together cause a resistance to rotation called the *moment of inertia*. Given three wheels of equal mass, the wheels with a larger diameter (and mass further from the rotation axis) will have a larger moment of inertia, which is why it is easier to travel long distances by bicycle than by rollerblades: the tiny, light wheels of rollerblades lose their angular momentum to friction much more quickly than the big, heavier wheels of a bicycle. In terms of the moment of inertia, the angular momentum of a spinning object is "angular momentum = (moment of inertia) times (rotations per minute)."

Angular momentum is, as we have said, a conserved quantity. If we start with a wheel at rest and spin it clockwise, something else must start spinning counterclockwise in order for the net angular momentum to remain zero. A good demonstration of this can be

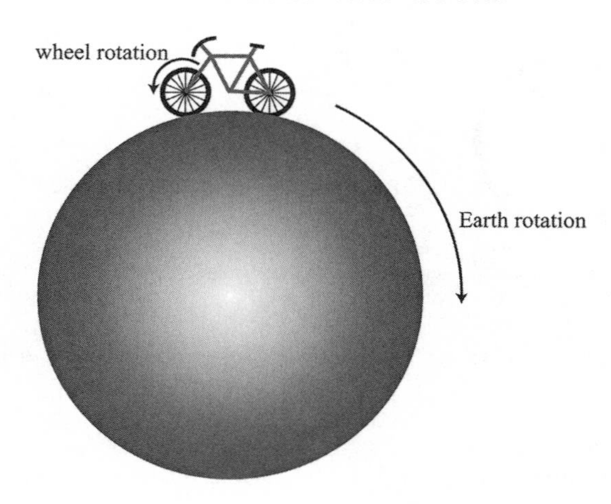

When the wheels of a bicycle spin, touching the Earth, the Earth spins a very small amount underneath them in the opposite direction. My drawing.

done in an ordinary spinnable office chair: if you twist your body suddenly to the left, the chair will twist to the right; angular momentum is conserved.

This conservation of angular momentum has some strange implications. When a bicyclist pedals, the angular momentum imparted to the wheels is countered by an equal and opposite angular momentum imparted to the Earth itself: the bicyclist makes the Earth rotate slightly! The Earth is so large and massive (it has such a high moment of inertia) that the actual rotation imparted to the Earth is negligible. Furthermore, there are so many people riding bicycles in different directions all over the Earth that, on average, all those tiny rotations effectively cancel each other out.

The idea of angular momentum was already hinted at in Newton's *Principia*, where he at least recognized a sort of "inertia of rotation," analogous to the inertia described in his First Law of Motion. As he states in his introduction to the First Law: "A top, whose parts by their cohesion are perpetually drawn aside from rectilinear motions, does not cease its rotation, otherwise than as it is retarded by the air. The greater bodies of the planets and comets, meeting with less resistance in more free spaces, preserve their motions both progressive and circular for a much longer time."[3]

Before Newton, the astronomer Johannes Kepler demonstrated through planetary observations what became known as the *Theorem of Areas*, according to which a given planet moves at a faster velocity when it is orbiting close to the Sun and at a slower velocity when it is orbiting farther away. Recalling that "angular momentum = (radius) times (mass) times (velocity)," we can conclude that if angular momentum is conserved but the radius shrinks, then velocity must increase. Over the next century and a half after Newton, a number of physicists gradually realized that the Theorem of Areas implied that there was some sort of general law for the conservation of rotational momentum. By 1800 there was a broad sense that the "momentum of rotation" is conserved; the term *angular momentum* was finally introduced by William J. M. Rankine in his 1858 *Manual of Applied Mechanics*.[4]

One simple—but, as we will see, misleading—implication of the preceding discussion is that angular momentum and rotation are directly related to one another. When we consider the case of an object like a bicycle wheel, we may say that it has angular momentum if it is rotating and no angular momentum if it is not rotating. Because angular momentum is conserved, this seems to imply that an object that starts out with no rotation cannot spontaneously start rotating. In particular, a cat that begins falling without any initial rotation could not, according to this reasoning, turn itself over, because this would imply that it had violated the law of conservation of angular momentum. Though cat enthusiasts like William Gordon Stables were continuing to use Antoine Parent's explanation of the falling cat from the year 1700, physicists of the late 1800s had recognized that Parent was wrong: a cat would not spontaneously flip over. They concluded that a cat must, at the moment it starts to fall, push off from whatever solid object is nearby—the ledge it is falling from, the hand that is dropping it—to gain some angular momentum to turn.

Marey's photographs showed clearly that this accepted hypothesis wasn't true. A description of the heated discussion about it appeared in the French magazine *La Joie de la Maison* under the title "The Hotbed of Scientific Thought!"[5] It is included here in full.

> The Academy of Sciences devoted one of its recent colloquia to the curious question of why a cat, like so many men of politics, always lands on its feet.
>
> This question, arising as it does from a selection of natural phenomena that discourage by dint of remaining inexplicable, gave the Academy quite a thrill.
>
> Other examples of these phenomena include the thorny problem of why chickens have no teeth; or why, when two men—one tall, the other small—meet in passing on a rainy day, it is always the smaller who attempts to lift his umbrella above that of the taller.
>
> In such a spirit of inquiry, Mr. Marey submitted to the learned scrutiny of the Academy sixty photographic images of a cat dropped from a height of five feet. The images show the cat falling with paws in the air, then twisting itself in time to alight on all four limbs.
>
> "Quite a marvel!" you might say—knowing, as the entire world has for some time, that when a cat falls, it always lands paws first.

> Everyone knows this, certainly; but the layman contents himself simply with knowing it, and lacks any desire for deeper examination. Mr. Marey, with the Academy by his side, demanded more. Their question was posed in the following manner:
>
> "Why does a falling cat always land on its feet?"

The gleeful sarcasm of the writer practically leaps off the page, and with good reason, as the simple falling cat problem would send the prestigious Academy into turmoil.

> And with that, my dear scholars were loosed into the field. Mr. Marcel Deprez pointed out that a falling body cannot be rotated without the application of an external force. Others holding forth on the subject included Mr. Loewy, Director of the Observatory, who might be better off observing falling stars than falling cats; Mr. Maurice Lévy, Mine Inspector (note: not a Feline Inspector); and Mr. Bertrand, Permanent Secretary of the Academy of Sciences.

There is an excellent French pun in this passage: *mines* is French for "mines," while *minets* is French for "felines." The reader is warned to not think Mr. Lévy, "inspecteur des mines," is the "Inspector of Felines."

> All shared the opinion of Mr. Marcel Deprez: the cat, they insisted, rotates because it stabilizes itself against the hand that launches it into space.
>
> It was then that Mr. Marey proffered his instant photographs, taken during the plunge. These images clearly

Rear view of the gardener's falling cat, 1894. The series of images, like others by Marey, is read right to left, top to bottom. Wikimedia Commons.

> show that at the beginning of its descent, the cat remains in its original position. Only when it comes to its senses and understands that things are going downhill does it right itself.

Marey's rear-view images of the falling cat, illustrated here, show this most clearly. In the first three images on the top row to the right, for example, when the cat is just released, it is not rotating much, if at all. Its change in orientation happens rapidly but only really begins when the cat is truly in freefall, away from any object it might push off from.

The physicists at the meeting had a difficult time accepting the evidence of their own eyes, and at least one of them decided that something quite unusual must be occurring.

> Vexed, but far from beaten, Mr. Marcel Deprez suggested that the cat might modify the position of its intestines in order

> to displace its center of gravity. This prompted a tart inquiry from a reporter to Mr. Marey: "Do you propose to know what occurs inside the belly of a cat?"
>
> Imprudent utterance, as it awakened in Mr. Marey a curiosity of unhealthy proportions. Be sure that this scholar will attack the question with his usual indomitability, slitting open any number of cats' bellies to find out what is inside.

As we have seen, these not-quite-serious fears were unfounded, since Marey was against vivisection as a physiological tool.

> But the last word did not belong to any single academician. Committing themselves entirely to the cause, the Academy's members entreated Mr. Marey with one voice to repeat his photographic experiments . . . only this time, he was to tie a string to the cat. Mr. Marey assured them that he would not fail to do so.
>
> As you can see, one never grows bored at the Academy. But suppose that an Academy of Cats came together to suspend at the end of a string a scholar, or a man of politics, so as to observe the manner in which he fell? We would not hesitate to conclude that cats are an ingenious yet ferocious sort of animal.
>
> For my part, I know more than a few scholars or men of politics who would land belly first in the course of such an experiment, as their paunches would reach the earth before their paws. See, for example, Mr. . . . well, never mind. I meant to quote potbellies, not personalities.

> But I have drifted far from the original question, and this is not the place for philosophizing. It is much simpler to resolve the problem before us, and I will attempt to do so to the general satisfaction.
>
> Having surpassed Mr. Marey, Mr. Marcel Deprez, Mr. Loewy, and the entire Academy of Sciences, I know quite well why a falling cat always lands on its feet:
>
> Because it hurts less!

It was clearly a delight for the journalist at the meeting to see such a distinguished group of scientists stumped by the overly familiar, even proverbial falling cat. It is also striking that the members of the Academy were so invested in their existing explanation—but where did that explanation come from? A hint can be found in a retrospective article published nearly a decade later, which introduces the incorrect theory that cats push off from something as they fall and then discusses Marey's explanation.

> A shining light in the scientific world declared this to be contrary to common sense and the laws of mechanics. No animal, he asserted, falling freely through the air, could possibly turn round by its own unaided exertions. Such cogent reasons did he bring forward that his colleagues in the French Academy of Sciences were forced, reluctantly, to accept his views.[6]

The "shining light" in this case was the French astronomer and mathematician Charles-Eugène Delaunay (1816–1872), who is best known for performing detailed motion studies of the Moon and for becoming the director of the Paris Observatory in 1870. He wrote a book on the physics of motion, *Traité de Méchanique Rationnelle*, in which he implicitly argued against a cat's being able to turn over without a fulcrum.

> If we suppose, as we have already done, that an animate being is isolated in the middle of space, that no external force is applied to it, and that it be primitively immobile, not only will this animated being not be able to move its center of gravity, but it will not be possible for it to move around this point. Indeed, in whatever way he plays his muscles, he can only develop inner forces; the absence of any external force therefore results in the sum of the areas described, in projection on any plane passing through its center of gravity, by the vector rays emanating from this point, constantly retaining the same value: therefore this sum of the areas must remain constantly zero, since, first of all by virtue of our hypothesis, the animated being in question was originally immobile.[7]

Delaunay argues that the Theorem of Areas prohibits a living creature, isolated in space, from being able to "move around" its center of gravity—that is, rotate. He does not explicitly mention cats, but the falling cat would be the prime example to which his argument would apply. As we will see, this argument is wrong. But Delaunay was held in such high esteem by his colleagues that they evidently did not look too deeply into the problem until Marey's photographs were right in front of their noses.

Delaunay was not the only prominent physicist who was on record for supporting the same argument. The French Academy could also point to our old friend James Clerk Maxwell to bolster their case.

As we have seen, Maxwell was involved in casual cat-turning experiments back in the 1850s, though he never formally published any hypotheses or observations in any scientific journal. However, he had apparently discussed his ideas at length with his lifelong friend and fellow Scottish physicist Peter Guthrie Tait. When Maxwell died unexpectedly in 1879, at the age of forty-eight, Tait was asked to write a scientific obituary for him in the pages of the prestigious journal *Nature*. On top of mentioning Maxwell's impressive accomplishments in electromagnetism, thermodynamics, astronomy, color, and mechanics, Tait felt compelled to include Maxwell's private thoughts on cat-turning.

> In his undergraduate days he made an experiment which, though to a certain extent physiological, was closely connected with physics. Its object was to determine why a cat always lights on its feet, however it may be let fall. He satisfied himself, by pitching a cat gently on a mattress stretched on the floor, giving it different initial amounts of rotation, that it instinctively made use of the conservation of Moment of Momentum, by stretching out its body if it were rotating so fast as otherwise to fall head foremost, and by drawing itself together if it were rotating too slowly.[8]

Tait's description answers one question as to how the explanation of a cat pushing off of a fixed body could possibly work. How could the cat know how hard to push in order to rotate the right amount? The answer, according to Maxwell, is that it doesn't: the cat uses its legs to adjust the speed of rotation so that it lands on its feet. This would work much the way an ice skater can adjust the speed of a spin on the ice: by drawing in the arms to spin faster and by extending the arms to spin slower. Similarly, Maxwell reasoned, the cat can adjust how quickly it is rotating by extending or retracting its legs, changing its moment of inertia to control its rotational speed.

This explanation was also incorrect, as Marey's photographs showed. In Maxwell's defense, he never published his idea himself, which suggests that he did not consider it strong enough for presentation to the public. It appeared only at Tait's initiative. But the combined strength of the reputations of Maxwell and Delaunay was enough to firmly embed the "fulcrum" hypothesis in the minds of the French Academy, leading to the argument at the fateful meeting in October 1894.

Fortunately, after the initial shock of seeing Marey's "scientific paradox," the French academicians rallied. With a little time to think about the physics and the math, the humbled Academy members came prepared to their next gathering.

> At the next meeting, M. Maurice Lévy rose and said that, in his opinion, the whole difficulty in the case had sprung from an inexact interpretation of some fundamental principles of mechanics. He then passed to the blackboard, which he rapidly covered with figures that proved clearly to the severest understandings that the cat broke no mathematical laws by its fall. Peace settled down on the Academy: *Causa finita est. [The case is settled.]*[9]

Lévy described the source of confusion accurately. The physicists had fallen prey to the often-quoted problem that "a little knowledge is a dangerous thing." In this case, the problem is that all of the physicists in question had—as we have throughout this chapter—considered rotating bodies that are more or less rigid: arms may extend or retract, but bends and twists of the bodies were not taken

My cat Cookie helpfully demonstrating the non-rigidity of the feline form.

into account. Their superficial understanding of rotating bodies, and the conservation of angular momentum, was based on such rigid rotators. Cats cannot in any sense of the term be considered rigid bodies.

A new hypothesis of cat-turning that apparently satisfied most of the Academy members at the time was put forth by the French mathematician Émile Guyou, who is better recognized for developing a particular projection of the Earth onto a flat map, now called the Guyou projection.[10] To understand Guyou's cat hypothesis, we first recall the example of a person twisting while sitting in an office chair. Much like a figure skater, a person in a chair can control how much the chair twists by having the arms out or in during the upper body twist. If the person's arms are out, the person has a high moment of inertia and the chair will countertwist significantly. If the person's arms are in, the person has a low moment of inertia, and the chair will countertwist relatively little. By adjusting the personal moment of inertia relative to the chair's, the sitter can control how much—or how little—the chair counterrotates.

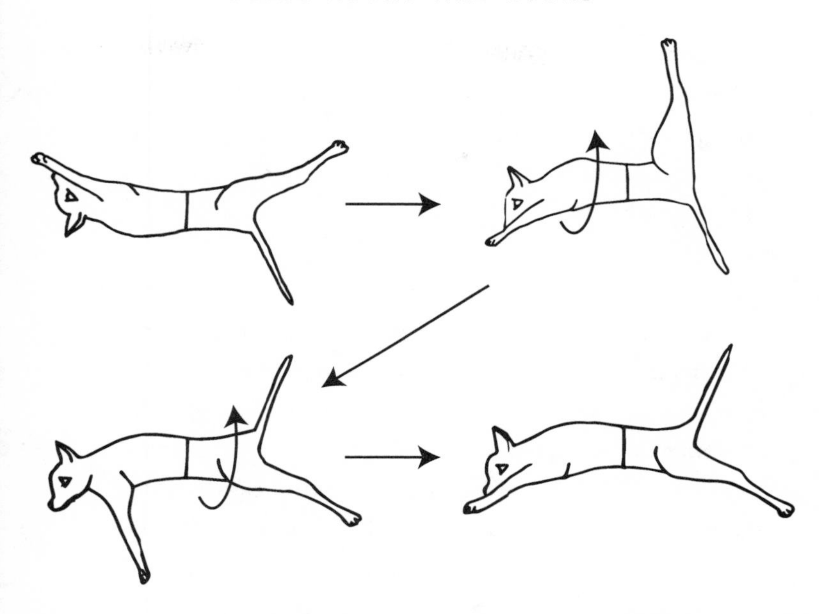

The tuck-and-turn model of cat-turning. Drawing by Sarah Addy.

Similarly, Guyou imagined that a cat could use its front and rear paws to control the moment of inertia of the front and back body sections. On initially falling, Guyou reasoned, the cat would extend its rear paws and tuck in its front paws. Then it could twist its upper body to be oriented correctly to the ground without significant countertwist from the lower body. Next the cat would tuck in its rear paws, extend its front paws, and twist the lower body to be correctly oriented without much countertwist from the upper body.

This explanation was endorsed by Marey, and it later became known as the tuck-and-turn model of cat-turning. Guyou's qualitative explanation was bolstered by Maurice Lévy, who showed with a rigorous mathematical analysis based on the Theorem of Areas that the explanation was, at the very least, physically plausible. Even Marcel Deprez, one of the most fervent defenders of Delaunay, came to accept the new view of the Theorem of Areas and, in an intriguing twist, revealed that he had been the one to motivate Marey's original cat photographs.

> The means used by the animal for this singular result seemed to have, from the point of view of the theorem cited above [Delaunay's theorem], great importance but, not having at my disposal any device for instant photography, I thought I could do no better than communicate my ideas on this subject to Mr. Marey, who possessed all means of investigation that I missed, and on several occasions, I expressed the desire to see him refer this matter to instant photography. I am pleased today by my insistence because the experience he has communicated to the Academy resulted, as I had told him, in drawing attention to the consequences of the theorem of areas and highlighting the error into which fell not only Delaunay but all the authors of treatises on theoretical mechanics.[11]

Both Guyou's and Lévy's explanations appeared right after Marey's paper on his photographs appeared in the journal *Comptes Rendus*; the academicians were presumably in a hurry to vindicate themselves after the ridicule they had received in the media. Even scientific publications had a little bit of fun with the whole affair; the account of Marey's photos in the journal *Nature* included an analysis from the cat's perspective: "The expression of offended dignity shown by the cat at the end of the first series indicates a want of interest in scientific investigation."[12]

What was not clear in the Academy's discussion, however, was whether Guyou's method describes the way that cats *actually* turn themselves over. Nevertheless, the explanation was accepted as fact, and within a few years it had even made it into physics books. An 1897 book on the "dynamics of a system of rigid bodies" includes this example.

> Ex. 3. Explain how a cat held with its feet upwards and let go is found, after falling through a sufficient height, to alight on its feet.
>
> During the first stage of the fall the cat stretches out its hind legs almost perpendicularly to the axis of the body and pulls the fore legs close to the neck. In this position it twists the fore part of the body through as large an angle as it can, the hinder part turning through a smaller angle in the opposite direction, so that the whole area conserved about the axis is zero. . . . In the second phase of the fall the attitude of the feet is reversed, the hind legs being close to the body and the fore legs pushed out. The cat now turns the hind part of the body through the large angle, the fore part rotating through the small angle. The result is that both parts of the cat are turned round the axis through nearly equal angles.[13]

Though the problem had seemingly been solved, one should not overestimate the impact that the revelation of the falling cat had on the thinking of physicists at the time. Using only a small number of photographs, Marey had demonstrated two important truths. The first is that although the laws of physics, such as conservation of angular momentum, cannot be broken, they can be "bent" in surprising ways that allow things to occur that superficially seem impossible.

The second of these truths is that nature has found many of these "bends" already and that a closer inspection of nature can potentially solve problems that are daunting to scientists.

Papers on the falling cat problem continued to appear in the *Comptes Rendus* for several months after Marey's work was published, further elaborating upon its implications. One of particular note was written by Léon Lecornu late in 1894. In it he noted that other animals might have different strategies for turning over.

> An even simpler example of inversion due to internal forces would be provided by a snake, which would be bent in a plane in the form of a torus and whose cross-sections would rotate, each in its plane, with the same angular velocity. The area theorem would obviously be respected, and not only would the outer shape be permanent but it would appear motionless in space, even while the back would come to take the place of the belly, and vice versa. It is not impossible that such processes may be used by some aquatic animals.[14]

Basically, Lecornu imagined a snake making a circle of itself like a donut or an Ouroboros and then rolling itself around its central axis. The angular momentum from the rotation of a segment of the snake's body on one side of the donut is countered by the angular momentum from the segment on the other side of the snake's body rotating in the opposite direction.

Lecornu's model was not taken as depicting a serious motion of a snake, though we will see that it is, ironically, closer to how a cat performs its feat than Guyou's model. It is of incidental interest to note that there is a Southeast Asian genus of snake, *Chrysopelea*, called the flying snake because of its ability and inclination to glide from the

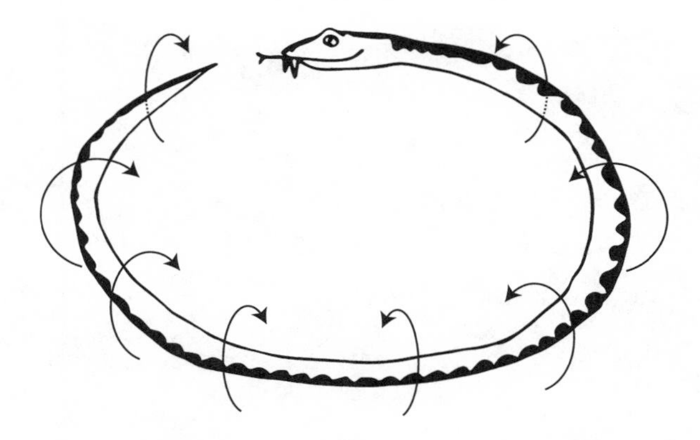

Léon Lecornu's model of a snake turning over in freefall. Drawing by Sarah Addy.

branches of trees to the ground. It achieves gliding by flattening its abdomen and undulating its body. Videos of the flying snake are well worth hunting down.

With the question of the falling cat solved to everyone's satisfaction at least for the moment, it seems that Marey did not follow through with his promise to tie a string to a cat before dropping it. He did, however, test chickens, rabbits, and puppies to see if they had a righting reflex. The dogs and the chickens failed, but, surprisingly, rabbits were able to flip over in a manner that on film looks remarkably similar to a cat's.

The revelations of the falling cat cemented the fame of Marey. Though he did not continue cat studies, he persisted with motion studies, and endless accolades and honors followed. In his research, he studied the mechanics of speech, the dynamics of bicycle riders, and the exertions of Olympic athletes. He became a consultant on various investigations into human flight, built a wind tunnel to study the motion of air currents around objects with his high-speed cameras,

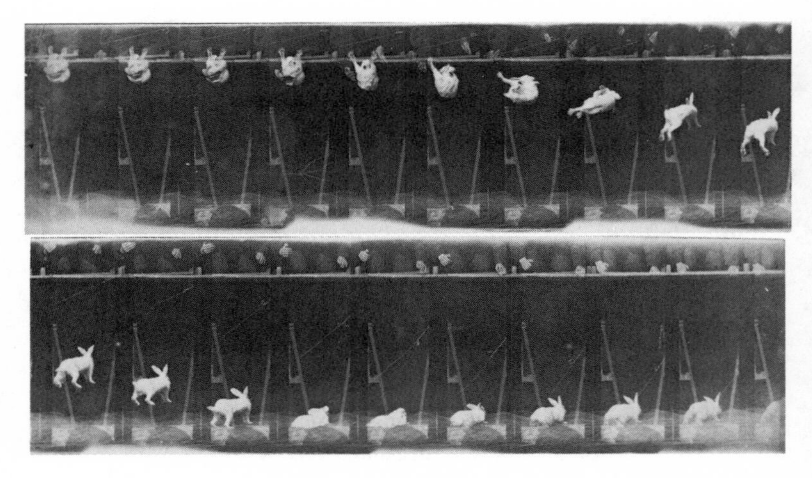

Marey's falling rabbits, shown in his "Lapin–Évolution de la chute," 1894. Albumen print; 4 × 23 9/16 in. (10.16 × 59.85 cm). San Francisco Museum of Modern Art, Purchase through a gift of Wes and Kate Mitchell and the Accessions Committee Fund. Photograph by Don Ross. Copyright © SFMOMA.

and looked at the motion of fluids. Around the time that he was presenting his research on falling cats to the French Academy, he was corresponding with both Thomas Edison and the Lumière brothers, all pioneers of cinema. Marey's and Demeny's work on projecting animated versions of their motion studies inspired Edison's Kinetoscope and the Lumière brothers' Cinématographe, the first commercial motion picture devices. Though Marey himself was not interested in motion pictures for entertainment purposes, his painstaking research paved the way for the titanic industry we know today.

In 1902, Marey's fifty-year association with the Collège de France was commemorated with a ceremony and the presentation of a medal, one side of which showed Marey in profile and the other side, Marey at work in his lab. By the end of that year, funds had been raised to complete the Institut Marey, a physiological institute that an international association of physiologists unanimously agreed to name after him.

When the Wright brothers achieved the first heavier-than-air powered flight in Kitty Hawk, North Carolina, on December 17, 1903, some small credit could be traced directly back to Marey again. In 1901, before the famous flight, Wilbur Wright had presented a paper to the Western Society of Engineers in Dayton, Ohio, in which he cited Marey.

> My own active interest in aeronautical problems dates back to the death of Lilienthal in 1896. The brief notice of his death which appeared in the telegraphic news at that time aroused a passive interest which had existed from my childhood, and led me to take down from the shelves of our home library a book on "Animal Mechanism," by Prof. Marey, which I had already read several times. From this I was led to read more modern works, and as my brother soon became equally interested with myself, we soon passed from the reading to the thinking, and finally to the working stage.[15]

It is unclear whether Marey was aware of the Wright brothers' groundbreaking achievement, since he was quite ill by late 1903. On May 15, 1904, he passed away from what is thought to have been liver cancer.

Muybridge's later career was somewhat rockier than Marey's. At first, things went spectacularly well: on November 26, 1881, ten days after Marey gave Muybridge an introduction to Paris science and society, another event was held in Muybridge's honor. This time, the host was the artist Meissonier, whose artistic sensibilities had originally been shocked by the photographs of the horse in motion; now, however, he welcomed the beauty and revelations of the photographs. When Muybridge showed his series of images, he was once again received rapturously: "The greatest applause followed the exhibition of each, and the many artists, whose greatest works on canvas or marble are those of the human figure, warmly congratulated Mr. Muybridge on the wonderful discovery which is destined to render such valuable aid to science and the arts."[16]

In letters, Muybridge's ambition seemed limitless: he was planning a joint collaboration with Meissonier, Marey, and an unnamed "capitalist" to produce the definitive text on animal motion. In 1882, Muybridge was invited to present a monograph on his photographic studies to the Royal Society of London; it would then be printed in their proceedings—an achievement that would cement his reputation as a man of science. However, three days before his presentation, the invitation was withdrawn. The Royal Society had received a book on animal motion photography titled *The Horse in Motion* by J. D. B. Stillman, "Executed and published under the auspices of Leland Stanford." The book gave almost no credit to Muybridge for his work, except as a "very skilled photographer." It seems that the egotistical Stanford, after Muybridge's departure, had hired Stillman to continue the photographic work and collaborate on the text.

The book undercut Muybridge's claim to being the original researcher; in 1883 he sued Stanford for injuring his professional reputation. Unsurprisingly, considering that most of the possible witnesses to the original work were Stanford's employees, Muybridge lost the case.

He was a resourceful man, however, and not easily kept down. In 1883 he entered into an agreement with the University of Pennsylvania to produce more motion studies, though these were largely of artistic interest: nudes engaging in mundane physical, sometimes erotic, activities. Some years later, in 1888, Muybridge met with Thomas Edison to discuss motion pictures and thus has a small claim to involvement in their development. But his great ideas, and his influence as a force to be reckoned with and listened to in photography, were behind him. At the 1893 World's Columbian Exposition in Chicago, Muybridge installed a "Zoopraxigraphical Hall" to show his looping motion studies, but the venture was a financial failure.

In the last years of his life, Eadweard Muybridge—once Edward Muggeridge—returned to his childhood home of Kingston upon Thames, where he seems to have lived a quiet life with his relatives, who were no doubt a little baffled by the eccentric and energetic man. He passed away on May 8, 1904, only one week before Marey. The two men, who were born the same year, died the same year, and had the same initials—Edward James Muggeridge and Étienne-Jules Marey—both had a monumental effect on motion pictures and photography as a whole.

Their passing marked the end of an era for photography, but the interest in falling cats had just begun. The cats that had caused such a stir at the French Academy had much more mischief to undertake.

6
Cats Rock the World

Marey's discovery that cats, and by extension any non-rigid bodies, can change their orientation in space without the need for angular momentum would have implications for a number of scientific fields. But the most immediate place that Marey's photographs had influence was in geophysics, where it inspired insights into the way the Earth spins. This influence also, however, sparked an infamous and long-lasting argument between two of the most important mathematicians of the late nineteenth century, Giuseppe Peano and Vito Volterra, in which the humble gardener's cat would play a prominent role.

The beginning of this very public fight can be traced to a paper by Peano that appeared in the January 1895 issue of the Italian journal *Rivista di Matematica*, with the title "The Principle of the Areas and the Story of a Cat."[1] (The principle of the areas is the Theorem of Areas.) Peano begins by summarizing a chaotic meeting at the Paris Academy and the explanations of the falling cat given by the attendees. He then gives his new explanation for how a cat achieves its incredible feat.

> But the explanation of the motion of the cat seems to me very simple. This animal, left to itself, with its tail describes a circle in the plane perpendicular to the axis of its body. As a result,

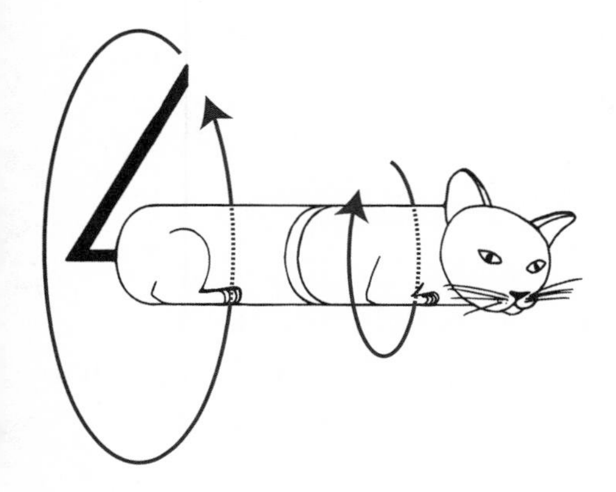

Giuseppe Peano's explanation of a turning cat. Drawing by Sarah Addy.

> through the principle of the areas, the rest of its body must rotate in the direction opposite to the motion of the tail, and when it has rotated the desired amount, it stops its tail and this simultaneously stops its rotatory motion while saving the essence and the principle of the areas.

In short, Peano suggests that if a cat spins its tail like a propeller in one direction, its body must spin in the opposite direction.

A cat's tail, however, weighs much less than the cat itself, which means that the tail would have to spin around more than once in order to flip the body over entirely. Peano seems to have realized this himself, for he notes that the cat may also swing its rear paws around to help with the motion.

> This movement of the tail can be seen very well with the naked eye and is equally clear from the photographs made. In these it is seen that the front paws, approaching the axis of rotation, do not affect the movement. The rear paws, stretched out near

> the revolution axis, perhaps describe cones in the same direction as the tail, and thus contribute to the body's rotation in the opposite direction. It follows that a tailless cat would turn over with much greater difficulty. Important note: do these experiences with a trusted cat!

Peano's argument is very much analogous to the office chair explanation of angular momentum conservation; in fact, he almost exactly describes this idea at the end of his paper.

> And if you swing a long stick in a horizontal plane, your body will rotate in the opposite direction. This stick corresponds to the cat's tail.

Peano's explanation is simple and elegant—too much so: nearly a century later, in 1989, J. E. Fredrickson would demonstrate experimentally that a tailless cat can turn over just fine, though cats possessing tails will use them to aid in the process.[2] But the propeller tail explanation was very much in character for a mathematician of Peano's style, strengths, and interests.

Giuseppe Peano (1858–1932), a formidable researcher in mathematics, published over two hundred books and papers. He was raised on a farm in the Italian village of Spinetta, and his earliest education was at the village school, where in cold months he brought a piece of wood from home to help heat the building during his classes.[3] Peano excelled at his studies, and his early brilliance was recognized by his uncle, who around 1870 invited him to stay with him and study in Turin. There, Peano attended a prominent high school; upon graduation in 1876, he enrolled in the University of Turin, where he would spend the rest of his career. After graduating from the university in

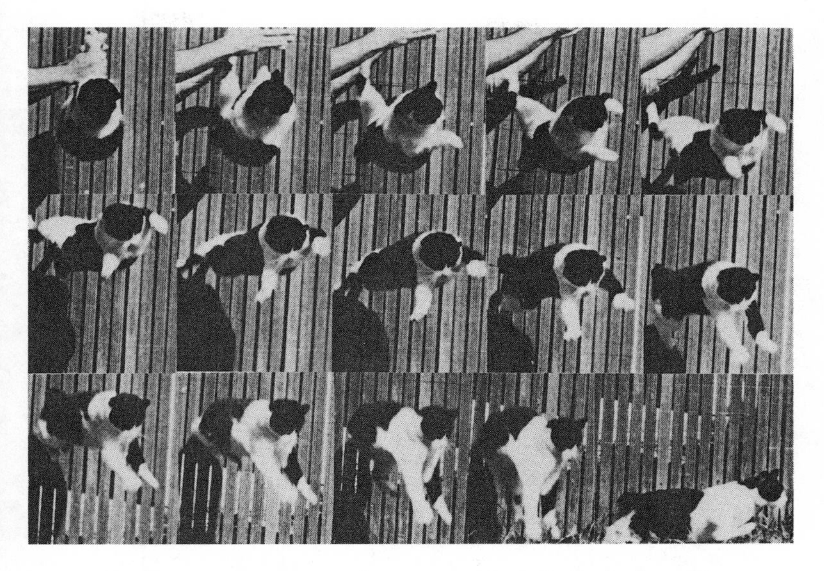

A tailless cat turning over, demonstrated by J. E. Fredrickson. From Fredrickson, "The Tail-Less Cat in Free-Fall," Physics Teacher, *27:620–625, 1989, reproduced with the permission of the American Association of Physics Teachers.*

Giuseppe Peano, circa 1910. Wikimedia Commons.

1880, he became an assistant to Angelo Genocchi, the chair of calculus, and was authorized to teach as well as begin his own mathematical researches.

It is during Peano's time under Genocchi that we first see some foreshadowing of future clashes to come. Peano seems to have been eager to make a name for himself. In 1882, for instance, he made his first significant mathematical discovery: an error in an important formula that appeared in a widely used calculus textbook. Peano wanted to publish the corrected formula but learned from Genocchi that the error and its correction had been found two years earlier, though not published. There followed some correspondence between Peano, Genocchi, and the original discoverer, Hermann Schwarz, as well as other mathematicians, for several years without much effect. When the first published announcement at last came out in 1890, it was by the ambitious Peano, not Schwarz.

Another example involved a more direct clash between Genocchi and Peano. Genocchi's lectures on calculus were highly regarded, and in 1883 Peano urged the senior mathematician to compile them into a book. On account of ill health, Genocchi declined, but Peano offered to write the book himself, in Genocchi's name. The book, *Calcolo differenziale e principii di calcolo integrale*, by Angelo Genocchi, appeared near the end of 1884, "published with additions by Dr. Giuseppe Peano."

The volume was, at least at first, a minor scandal. Peano not only compiled Genocchi's lectures but included in the book what he himself referred to as "important additions." This turn of phrase came across as both egotistical and disrespectful to the title author. How could a young upstart improve upon the work of the master? Genocchi himself was initially angered, though he seems to have eventually come around to appreciate the book as a whole. In hindsight, the additions *were* very important.

In spite of Peano's brazen approach to self-promotion—or likely in part because of it—he rapidly rose in rank and importance. In 1886 he took a second appointment as a professor at the Royal Military Academy, and in 1890 he earned a full professorship at Turin. It was during this time that Peano published some of his most interesting and important work. One of his grandest achievements was the formulation of what are now called the Peano axioms, a small simple set of statements describing all the properties of the natural numbers (0, 1, 2, 3, . . .). He also was the adopter and champion of a formal and standardized "language" that could be used to describe mathematical statements. This language allows often-lengthy mathematical proofs to be dramatically abbreviated. Peano's notation is still used in very much the same form to this day. In 1890 he co-founded the journal *Rivista di Matematica*, in which he published his first cat paper, "The Principle of the Areas and the Story of a Cat," and in 1891 he started the "Formulario Project," whose goal was to create a standardized encyclopedia of mathematics using the symbolic language he had developed.

One other example of Peano's work is worth noting here: the concept of a space-filling curve. The idea may be introduced by asking, Is it possible to draw a single line that fills in a square completely? With a pen and paper, we can always fill in a square, because the pen tip has a finite width. In mathematics, however, a line is an object with length but not width, whereas a square has length and width. Intuitively, we would expect that, in this sense, a square is "bigger" than a line. We often discuss this by talking about the *dimensions* of the objects: a line is a one-dimensional object; a square is two-dimensional.

By the late 1800s, advances in mathematics had demonstrated that the number of mathematical points in a line and a square are *exactly the same*. In principle, then, it should be possible to fill a square with a single continuous line; it was Peano who first showed explicitly how to do so. The construction he used is shown in the nearby figure, where the square is filled by a line traversing an increasingly twisted path. The first path is simply an S-shape. The next iteration involves taking S-shaped detours along that original path, the iteration after that, taking detours along the detours, and so on. Peano was able to rigorously prove that doing this iteration an infinite number of times results in a single unbroken line that passes through every point of the square—in fact, passes through each point multiple times.[4]

Much later, it would be recognized that Peano had discovered a very curious example of a mathematical object, a *fractal*. Ordinary geometrical objects have a dimension given by a non-fractional number—a square is two-dimensional, whereas a line is one-dimensional—and this number is in a sense a measure of how much space the object takes up. Fractals are objects that have a fractional dimension, indicating that the amount of space taken up by a fractal is drastically different from the amount of space taken up by simple objects. A fractal of dimension 1.5, for example, takes up more space than a line but less than a square. Fractals are often described as looking essentially the same at any level of magnification, much the way a thin section of a tree branch looks similar to a thick section of a tree branch. This self-similarity is also present in Peano's curve. In his unusual construction, Peano had found an odd fractal with a fractal dimension equal to 2—a non-fractional fractal.

As we have seen, Peano was an ambitious and imaginative mathematician who was typically interested in big-picture projects.

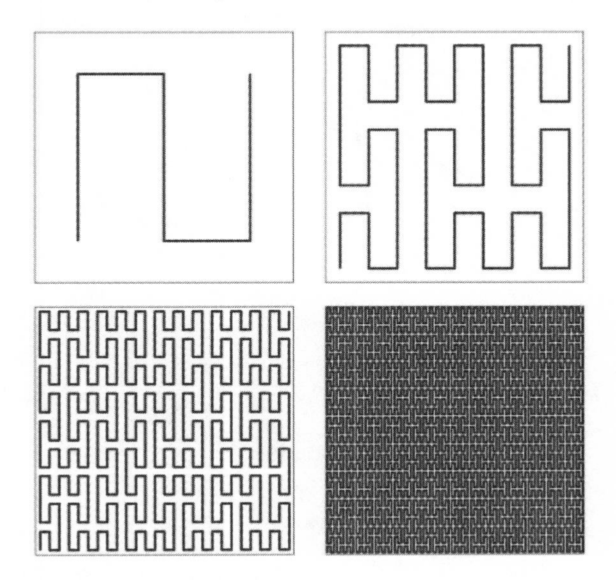

Four iterations of the Peano curve. My drawing.

However, he was also keen to demonstrate that the big-picture mathematical tools he was using were applicable to real-world problems. After thinking about the problem of the falling cat for a while, he saw in it an explanation for a problem in geophysics that was of great interest at the time: the *Chandler wobble*.

Already by Peano's time, astronomers realized that the direction of the Earth's axis of rotation is not fixed. As with a spinning top or gyroscope, the axis traces out a circular path, a *precession*. The Earth's precession has a period of 26,000 years. The axis wobbles slightly in that path, or undergoes a *nutation*, with a period of 18.6 years. The precession and the nutation are driven by the interaction of the Earth with the gravitational forces of the Sun and the Moon.

Another form of nutation was predicted in 1765 by the mathematician Leonhard Euler, who suggested that the spheroidal (slightly

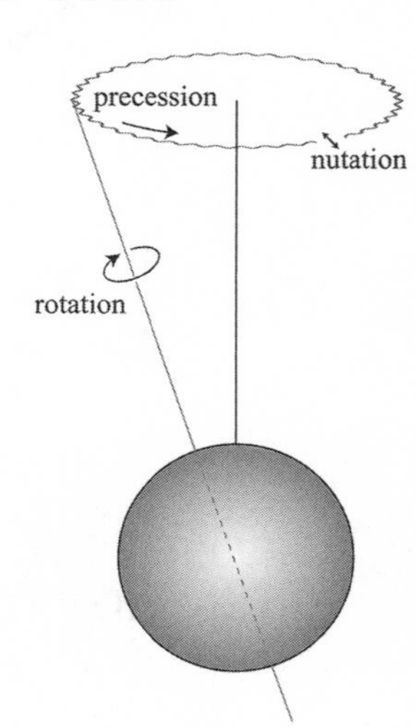

Illustration of precession and nutation. My drawing.

non-spherical) shape of the Earth allows for a "free nutation": an additional small wobble of the Earth's axis with respect to the solid Earth that is self-contained and not induced by external forces. This wobble arises because the axis of symmetry of the Earth—the axis with respect to which the Earth looks symmetrical—is slightly different from the axis of rotation, the axis around which the Earth actually turns. Executing some remarkable mathematical gymnastics, Euler predicted that this free nutation should have a period of 306 days.

This wobble, expected to be an exceedingly small variation in the direction of the Earth's axis, would require careful measurements of the positions of stars, as seen from the Earth, over at least a year to be detected. Such formidable obstacles are seen as a challenge by scientists, and for more than a century, numerous researchers attempted to observe Euler's predicted free nutation. They met with no success, and by the 1880s, astronomers had basically given up on searching for the effect.

It was about this time that Seth Carlo Chandler, Jr. (1846–1913), life insurance actuary and amateur astronomer, serendipitously found the very phenomenon that had eluded so many professionals.[5] Born in Boston, Massachusetts, Chandler received an early impetus in science when in his last year of high school he was given a job working for the Harvard mathematician Benjamin Pierce. Pierce, who collaborated with colleagues at the Harvard College Observatory, put Chandler to work doing mathematical calculations. After Chandler graduated, his skills earned him a job at the U.S. Coast Survey making astronomical measurements of longitude and latitude. After his supervisor left the survey, Chandler went into the insurance business, but his real love was astronomy; thanks to his Harvard connections, he was able to make measurements at the Harvard Observatory.

To measure latitude, Chandler had used a visual zenith telescope, which was designed to be pointed straight up into the sky; latitude could be determined by measuring the relative positions of the stars. During his employment with the Coast Survey, Chandler had noted that it took a lot of effort to properly level the telescope, almost doubling the amount of time needed to take measurements. For his first project as an amateur astronomer, then, he designed a new device that could level itself, which he called an Almucantar. From mid-1884 to mid-1885, Chandler tested the accuracy of the Almucantar at the Harvard Observatory; his measurements revealed, unexpectedly, that there was apparently a continuous systematic change in the latitude of the Observatory over the course of a year. These were his first measurements of the wobble. Chandler himself did not speculate on their origin; he only noted that he could find no source of error that could account for the observations.

This issue might have remained dormant for many more years except for a remarkable coincidence: over almost the same period of time that Chandler had done his work, the German scientist Friedrich Küstner of the Berlin Observatory had also observed a variation of latitude. Küstner had, like Chandler, been attempting to study something else entirely—in Küstner's case, variations in the speed of light coming from distant stars. This effort was later shown to be doomed, for Einstein's special theory of relativity reveals that the speed of light is the same for anyone who cares to measure it. Unsurprisingly, Küstner found no variations in light speed, and he could not explain the measured variations in latitude; he ended up putting aside his work for nearly two years. When he finally got around to publishing his results in 1888, he may have been motivated to do so by seeing Chandler's work.

Chandler, in turn, saw Küstner's results and realized that the change in latitude that he had measured was a real effect. He redoubled his efforts with the Almucantar and in 1891 published his first two papers on the Chandler wobble, showing a variation of the axial position of the North Pole of about thirty feet (nine meters), with a period of 427 days.[6]

Chandler seems to have discovered this variation where others had failed simply because he didn't know what he was looking for. Earlier astronomers hunting for the wobble had focused on Euler's estimate of a 306-day period and discounted any longer-period variations as seasonal changes in the atmosphere, which could, in principle, change the apparent location of stars. But Chandler, not familiar with Euler's result, simply measured the data without any preconceived target.

Chandler's results were conclusive. He not only used his own set of extensive measurement data to prove the existence of the wobble but showed that Küstner's data were consistent with his own and, further, that measurements from observatories in Pulkovo, Russia, and in Washington, D.C., showed the same variation.

The reaction to Chandler's discovery matched the later arguments over Marey's cat photographs: initial disbelief and bafflement, followed by rapid excitement and acceptance. A report from the Seventy-Third Annual Meeting of the Royal Astronomical Society in February 1893 summarizes the response.

> Astronomers had hesitated to accept the 427-day period, even in face of the very strong evidence of the 1860–1880 observations, owing to the difficulty in accounting for it theoretically. It had been pointed out by Euler that, treating the Earth as a rigid body, the period of rotation of the pole must be 306 days. Professor Newcomb, however, happily pointed out that a qualified rigidity (either actual viscosity or the composite character due to the ocean) afforded an explanation of this longer period; and after this suggestion Mr. Chandler's 427-day period was well and even warmly received.[7]

In short, Euler had assumed the Earth to be a perfectly rigid body, but the presence of fluids on the exterior of the planet—the atmosphere and the oceans—could result in significant departures from Euler's calculation.

It is one thing to provide an explanation, in words, for a new physical phenomenon, and another thing entirely to provide a quantitative theory to back up the explanation. When Peano encountered the problem of the falling cat in 1894, he immediately saw it as kindred to the wobbling Earth problem and began working on mathematics to explain the latter. Both problems involve an object changing its orientation in space even in the absence of external forces, and both problems can be qualitatively explained by internal motions of the object in question.

It is ironic that Peano was inspired by the falling cat. Antoine Parent had, in 1700, modeled a cat as a sphere; in 1895 we find Peano modeling the spherical Earth as a cat. On May 5, 1895, in a paper titled "Concerning the Pole Shift of the Earth," Peano presented to the Academy of Science in Turin his own mathematical theory of the phenomenon, giving due acknowledgment to the cat.

> At the end of last year at the Academy of Sciences in Paris it was proven by experiment that certain animals, such as cats, can, as they fall, through internal actions, change their orientation. The possibility of this motion is immediately explained by mechanics. In a short article published in the *Rivista di Matematica* (in the beginning of January 1895), I briefly discussed the question. I tried to describe the cyclic motions by which the cat actually rights itself and added other examples.
>
> It leads naturally to the question: Can the globe change its orientation in space using only internal forces like every living thing? Mechanically, the question is the same. But to Professor Volterra is accorded the merit of having made the first proposal. He made it the subject of some notes presented to this Academy, of which the first was on February 3rd.

In the next paragraph of his paper, Peano demonstrates his ability to explain physical concepts in a clear and entertaining manner in summarizing the concepts of conservation of momentum and angular momentum.

> As it is stated that when a body falls on the earth, the earth is approaching the body, so it can be said that any displacement of a body on the earth produces an opposite movement of the globe. So if Mohammed goes to the mountain, the mountain approaches Mohammed; and if a horse makes a trip around a racecourse, it forces the land to rotate in the opposite direction, with the difference, however, that if the horse, after describing a circle, returns to the place from which it started, the earth has rotated only a very small angle and has assumed a different orientation from that as if the horse had not moved.

Simon Newcomb had already suggested that the non-rigidity of the Earth could explain the Chandler wobble; Peano's work was novel in that he evaluated specific mechanisms that could contribute to the wobble.

> On land the waters of the seas are moving in the form of currents; in the atmosphere the water rises in vapor form, carried by the wind, and falls as rain or snow and fertilizes the plains, and through the beds of rivers returns to the sea.
>
> The purpose of this note is to explain how we can make the calculation of displacements produced on the earth by the relative motion of its parts and make a numerical estimate.[8]

The Gulf Stream, for example, circulates counterclockwise, bringing warm water from the tropics up toward Europe. This continuous circulation of water, according to Peano, would cause a corresponding small clockwise rotation of the Earth to balance it because of the conservation of angular momentum. Peano, in essence, envisioned the circulating Gulf Stream as acting like the cat's tail in his explanation of the falling cat.

Almost lost in the introduction is the seemingly gracious acknowledgment of the work of Professor Volterra: "But to Professor Volterra is accorded the merit of having made the first proposal. He made it the subject of some notes presented to this Academy, of which the first was on February 3rd." In fact, Volterra had presented his own mathematical analysis of the Chandler wobble, also arguing that marine currents could be the cause of the anomalous period.[9] So it appeared that Peano was acknowledging that Volterra was the first to work on the subject, but what Peano had actually done was throw down a challenge, one that would enrage Volterra and spark a yearlong feud.

Vito Volterra (1860–1940), like Peano, came from humble beginnings and showed his brilliance at an early age.[10] Born in the seaport of Ancona, Italy, Vito was only two years old when his father died, leaving him and his mother in the care of her brother. They settled in Florence, where Volterra spent much of his youth.

Vito Volterra, circa 1910.
Wikimedia Commons.

Volterra showed an early interest in mathematics, studying classic books on arithmetic and geometry at age eleven. At age thirteen, he read Jules Verne's classic novel *Around the Moon* and was inspired to calculate the trajectory of a projectile moving in the combined gravitational field of the Earth and the Moon; four decades later, he would belatedly present his method of solution in a series of lectures. By age fourteen, he was studying calculus by himself, with no teacher.

Volterra's rather poor family wanted him to choose a financially lucrative career, so his insistence on going into science dismayed them. In desperation, they contacted a wealthy and successful cousin, Edoardo Almagià, and asked him to talk some sense into the young man. Almagià's conversation with Vito so impressed the financier, however, that he reversed course and wholeheartedly encouraged Vito to follow his dreams. Volterra started his studies at the University of Florence, then attended courses at the University of Pisa, graduating as a doctor of physics in 1882. At age twenty-three, he became a full professor at the University of Pisa; about a decade later, in 1892, he moved to the University of Turin, where Peano was already established, to become a professor of mechanics.

To understand why Volterra was enraged by Peano's superficially benign acknowledgment, we must look at the dates mentioned. Peano notes that his article on cat physics appeared in January 1895 but that Volterra's first presentation on the subject of the Chandler wobble appeared in February. In other words, Peano was implying that Volterra got his ideas on the origin of the Chandler wobble from Peano's cat paper. If we imagine the research on the Chandler wobble as an uninhabited island, Peano's statement was the equivalent of sticking a large flag into the ground and claiming the territory as his own—with Volterra already standing there.

There are many types of feuds in the scientific community, but none of them are typically more angry, and less productive, than a fight over the priority of discovery—that is, who found it first. Being the first to discover a phenomenon or an explanation of a phenomenon can make or break a career, and priority is, surprisingly often, decided by a difference of just a few weeks or even days. Volterra, for his part, was justified in his outrage; he had been working on the problem for longer than Peano was willing to acknowledge. Though his first paper on the subject appeared in the journal *Astronomische Nachrichten* in February 1895, he had submitted it for publication months earlier.[11] From Volterra's perspective, Peano was swooping in at the last moment to claim credit for something Volterra had been studying for a year; even worse, he was practically accusing Volterra of stealing the idea, without credit, from Peano's cat paper.

Volterra, who was present for Peano's talk on May 5, immediately objected. Volterra told the assembled Academy members that he had been working on the problem for quite some time but had held back more detailed calculations, based on Chandler's data, until he had time to properly study the data. With the Academy's permission, he would retrieve his paper to show his work. He was given this permission, and he did fetch his paper and present it.[12]

This volley began the war between the two mathematicians. As we saw earlier, Peano was not one to back away from a fight over priority. He presented another paper to the Academy of Turin on May 19; however, he found an error in his notes and withdrew the paper before publication. In the meantime, Volterra was busy, presenting two additional papers with more detailed calculations to the Turin Academy on June 9 and 23.[13]

Scientific researchers rely on other people to acknowledge their work. It is therefore considered good etiquette, when writing a new research paper, to include citations to all relevant publications that came before. In Volterra's two June papers, he referred only to his own work; he did not mention Peano's recent contributions at all. Peano seems to have noticed this; in his next paper, also presented on June 23, he mentions the work of everyone who had studied the latitude wobble before—Zanotti-Bianco, Eneström, Bessel, Gyldén, Resal, Thomson, Darwin, Schiaparelli—everyone except *Volterra*.[14]

After this, Volterra had enough of the bickering in Turin. He sent his next publications on the subject to the Accademia dei Lincei in Rome. In literal translation, it is the "Academy of the Lynx-Eyed," with the keen-eyed lynx as a symbol of the clear perception that scientific work requires. Maybe Volterra, plagued by Peano's cat, saw the irony in making the Lincei the next battleground in his war with Peano. In any case, the Accademia dei Lincei was a venerable institution. Founded in 1603, it had been revived in the 1870s to become the top scientific institution in Italy. Volterra was, in essence, appealing his case to the highest science authorities in the country.

Volterra published two papers in the journal of the Lincei. The first, "received before the 1st of September, 1895," is a general discussion of the mathematics used in his solution to the wobble question and appears to have been written to demonstrate his knowledge of the subject.[15] The second paper, received before September 15, calls out Peano directly.

> Professor Peano in a note presented to the Academy of Turin in the session of June 23 of this year, which has just now been printed, shows that a system which is symmetric about an axis

> and which constantly maintains its form and density distribution, may have variable internal movements that follow a law such that the rotational pole moves continually farther from the inertial pole.
>
> Seeing that this result can be obtained as an evident and immediate consequence of formulas considered by me and explained in several preceding memoirs, which Professor Peano forgot to cite, although they were published this year in the same Acts of the Academy of Turin, I may be allowed to show this here, avoiding the employment, made by said author, of methods and notations not generally accepted, and of proceedings hardly suited to making clear the path taken and the result reached.[16]

Here Volterra makes explicit his annoyance at not being cited by Peano. He also, however, alludes to a significant point of contention between the two authors: "methods and notations not generally accepted." Peano, we have seen, was an innovator of new mathematical techniques and new notations for those techniques. In the puzzle of the wobbling Earth, Peano saw an opportunity to prove that his novel methodology was of practical use. Volterra, in contrast, was very much in favor of traditional techniques in calculus and was happy to mock Peano's new ideas.

Peano was more than up to the challenge of responding to Volterra and had no problem with further antagonizing his opponent. In a paper dated December 1, 1895, Peano addresses Volterra's criticisms.

> In an article with this same title, and published in the Proceedings of the Accademia dei Lincei on September 15, Professor

> Volterra confirms with his calculations one of the results you find in my two Notes, "Sullo spostamento del polo terrestre," published in the Acts of the Academy of Turin, on May 5 and June 23. And since the issue of the pole's motion is now very interesting, I think it is useful to expound in a few words the results I came across, which can be found by either way.[17]

Peano seems to start with a psychological mind game: he gave his article the same title as Volterra's, but by explicitly noting this manages to make it sound as though Volterra had somehow stolen the title from him in advance. Furthermore, Peano says that Volterra "confirms" Peano's results, again suggesting that Volterra was simply following Peano's groundbreaking work.

He then returns to a discussion of the cat problem that started the whole argument.

> It is well known that about a year ago (October 29 and November 5, 1894), in the Academy of Sciences of Paris, the common statement was that a cat, however dropped, always falls on its paws. And if for a moment this was considered to be contrary to the principle of the areas, then it was easily recognized that this principle, properly understood, completely explains the phenomenon. I also briefly addressed the issue in the *Rivista di Matematica* (January 1895).
>
> . . .
>
> Discussing later the question of the earth's pole shift, produced by the motions of parts of the earth, such as ocean currents, I pointed out to some people the identity of the two issues, because instead of the cat and its tail one can speak of the earth and its ocean.

It seems quite likely that Peano vaguely wrote "some people" to let the reader imagine that Volterra himself was one of them and that Volterra got his explanation for the Chandler wobble from discussions with Peano. If true, or at least if believed by the majority of scientists, this would give Peano the priority of discovery.

Peano's article seems to have been the last straw for Volterra. In exasperation, he wrote a letter to the president of the Lincei, dated January 1, 1896, in which he blasts Peano.

> Dear Mr. President:
>
> Please allow me to communicate to you a brief reply to the note of Professor Peano. . . .
>
> Relative to what is found said in the beginning of his note, it seems to me that it is not worth the effort of spending any words, seeing that no one can doubt my priority, whether with respect to treating the question, or with respect to the fundamental idea which forms its point of departure; nor can any doubt arise about the originality of my idea, as I explained in my lectures of last year, which I found while searching for an apt example to illustrate Hertz' concept of substituting hidden movements for the consideration of forces in the investigation of natural phenomena; and it is not necessary for me to justify myself with the cat question, as Peano hints, a question, for that matter, about which he limited himself to writing in his journal a simple and brief review of the work of others. . . .
>
> Thus, having announced this conclusion, which was immediate and evident from my considerations, without citing me, but only dressing it up in vectorial language, Peano deserves the censure contained in the note I presented to the

> Academy last September. And so, I have no need of agreeing with any of Peano's results.
>
> . . .
>
> Having thus shown to be empty and unfounded any of the points of criticism made of me by Peano, and that his assertions are neither original nor exact, he himself having recognized them as such, for my part I hold this polemic definitively closed.[18]

Peano seems also to have had enough of the argument, though he had to get in the last word. In a final paper to the Lincei dated March 1, 1896, he re-derived some of his earlier formulas in a more explicit form, in essence "showing his work" so that readers would not doubt that he had achieved the results he had claimed to in his earlier papers.[19] Probably wisely, he did not mention Volterra at all, and with this quiet discussion the war between Peano and Volterra drew to a close.

It was a remarkably intense fight for what would seem to have been a relatively small scientific discovery. Between the two of them, Peano and Volterra published fourteen papers on the Chandler wobble over about a span of a year—an incredible rate of output on a single specific problem. Peano's motivation probably came, at least in part, from the opportunity he saw to demonstrate the practicality of the new mathematical techniques that he had championed. Volterra may have been motivated by Peano's formalism, but in the opposite direction. The two mathematicians both worked at the University of Turin, and Peano was pushing for all professors to use his new methods in the classroom. The traditionalist Volterra may have resented this and may have been genuinely enraged by Peano's attempt to push those methods into his research problems as well.

But what of the Chandler wobble itself? Though the broad explanation of ocean processes championed by both Peano and Volterra has held up through the years, a detailed understanding of the Chandler wobble has remained somewhat elusive. Early in the twentieth century, researchers found that the wobble is more complicated than Peano or Volterra ever imagined: for example, the size of the wobble can vary over the course of decades, occasionally making dramatic "jumps" in its behavior. There are also multiple sources of the wobble: in 2000, Richard Gross of the Jet Propulsion Laboratory at the California Institute of Technology showed through simulations that the dominant source of the wobble over the years 1985–1996 was fluctuations of pressure at the ocean's bottom, with other ocean and atmospheric effects playing smaller roles.[20]

So Peano and Volterra's fight was not as important as either of them had hoped. But the fight would not have occurred at all if not for a series of photographs of Étienne-Jules Marey's gardener's cat. In this case, at least, cats earned their reputation as creators of mischief.

7

The Cat-Righting Reflex

Marey's photographs of a falling cat shocked physicists and forced them to rethink their preconceived notions of how objects move and turn in space. But this was nothing compared to the seismic shock that traveled through the scientific community in 1905 and changed our understanding of physics forever. In that year, a then-obscure patent clerk named Albert Einstein published three papers in the German journal *Annalen der Physik*, each of which would serve as the foundation of a new branch of physics. This trio of papers is today referred to as the *annus mirabilis* (miracle year) papers.

The first, "On a Heuristic Viewpoint Concerning the Production and Transformation of Light," appeared on June 9. In it Einstein sought to explain the *photoelectric effect*: why light shining on a metal plate can cause the plate to eject electrons. Einstein argued that the effect could only be explained by considering light a stream of particles, even though it had already been demonstrated that light acts as a wave. *Wave-particle duality* is now a fundamental aspect of quantum physics. Einstein would, in 1921, win the Nobel Prize in Physics for his work on this subject.

The second of his 1905 papers, "On the Motion of Small Particles Suspended in a Stationary Liquid, as Required by the Molecular

Kinetic Theory of Heat," appeared on July 18 and explained the phenomenon of *Brownian motion*, the apparently random bouncing of small particles in hot water. Einstein showed that this peculiar motion could be explained as arising from collisions between the particles and the otherwise indiscernible water molecules surrounding them. The explanation led to the final confirmation that matter consists of discrete atoms and molecules; surprisingly, there was still some lingering doubt about this even at the beginning of the twentieth century.

The third of Einstein's 1905 papers, "On the Electrodynamics of Moving Bodies," appeared on September 26. It is the most celebrated of the three, for it was the first statement of Einstein's *special theory of relativity*, which would revolutionize our view of space and time. To understand the significance of this work, we need a little background.

One of the founding principles of physics, dating back to Galileo Galilei, is the *principle of relativity*, which can be simply stated as "the laws of physics are the same for any observer, regardless of the observer's motion." In a 1632 text, *Dialogue Concerning the Two Chief World Systems*, Galileo explains it thus:

> Shut yourself up with some friend in the main cabin below decks on some large ship, and have with you there some flies, butterflies, and other small flying animals. Have a large bowl of water with some fish in it; hang up a bottle that empties drop by drop into a wide vessel beneath it. With the ship standing still, observe carefully how the little animals fly with equal speed to all sides of the cabin. The fish swim indifferently in all directions; the drops fall into the vessel beneath; and, in throwing something to your friend, you need throw it no more strongly in one direction than another, the distances

> being equal; jumping with your feet together, you pass equal spaces in every direction. When you have observed all these things carefully (though doubtless when the ship is standing still everything must happen in this way), have the ship proceed with any speed you like, so long as the motion is uniform and not fluctuating this way and that. You will discover not the least change in all the effects named, nor could you tell from any of them whether the ship was moving or standing still. In jumping, you will pass on the floor the same spaces as before, nor will you make larger jumps toward the stern than toward the prow even though the ship is moving quite rapidly, despite the fact that during the time that you are in the air the floor under you will be going in a direction opposite to your jump. In throwing something to your companion, you will need no more force to get it to him whether he is in the direction of the bow or the stern, with yourself situated opposite. The droplets will fall as before into the vessel beneath without dropping toward the stern, although while the drops are in the air the ship runs many spans.[1]

Galileo realized that, when sitting within the depths of a ship, it is impossible to tell by any experiment whether the ship is at rest or moving at a constant speed; living creatures, whether they are walking, swimming, or flying, will be unable to detect any motion. Consider, for example, a game of tennis taking place inside a moving ship. Many people would assume that the tennis ball will have a tendency to drift toward the back of the ship as the ship moves forward, giving the player nearer the prow the advantage, but this intuition is incorrect. The ball will act in all ways as if the ship were docked unmoving in a harbor. If no physics experiment can determine the

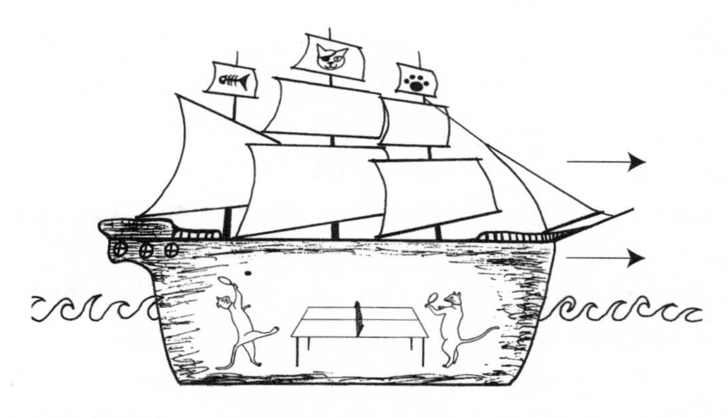

A game of "Galilean tennis." Even if the ship is moving to the right, neither player will have any advantage, contrary to the common and mistaken intuition that the ball will tend to move more toward the back of the ship. Drawing by Sarah Addy.

motion of the ship, the laws of physics must be the same for any observer moving at a constant speed.

Following Galileo, Isaac Newton successfully applied this principle to his famous laws of motion and codified relativity into the movement of any material bodies. For example, an observer standing next to a billiard table and an observer walking past it can both describe everything that happens in the game using Newton's laws, even though they will disagree on how fast the balls are moving relative to their own positions.

When James Clerk Maxwell argued that light is an electromagnetic wave in the 1860s, however, it was quickly recognized that Newton's form of relativity did not apply to the waves. In particular, Newton's formulas indicated that observers moving at different speeds would generally measure different values for the speed of light. A person moving parallel to a photon, for instance, would see it moving more slowly than would a person moving anti-parallel to the same photon. Since the speed of light is built into Maxwell's equations, the equations would be slightly different for each of these observers. Physicists concluded that the physics of light waves must work differently for each observer. From Maxwell's time to Einstein's, a number of experiments were performed to measure the presumed variations in the speed of light, all without success. The most famous of these was an 1887 experiment by Albert A. Michelson and Edward W. Morley, who employed the interference of light waves to measure changes in the speed of light; they were unable to detect any change, even though the ongoing motion of the Earth around the Sun should have produced a measurable effect.

Albert Einstein attacked the problem from the other direction. He asked: If the laws of electricity and magnetism are the same for any moving observer, what would the principle of relativity look like? He made two assumptions in his calculation: (1) *all* laws of physics are the same for all observers moving at constant speeds, and (2) the speed of light is the same for all observers. From these two assumptions, a bewildering array of bizarre consequences followed. Among them are:

- nothing (that we know of) can move faster than the speed of light
- mass and energy are equivalent, and one can be converted into the other (this is represented by the famous equation $E = mc^2$)
- time moves slower for a moving object
- moving objects have their length contracted along the direction of motion
- time and space are in a sense inseparable and form a four-dimensional entity known as *spacetime*

In the century since Einstein's publication, all of the strange predictions of special relativity have been confirmed in a wide variety of experiments.

Relativity may seem quite removed from our discussion of falling cats. But Einstein's next project would turn out to have great pertinence to the problem. Almost immediately after finding success with his special theory of relativity, Einstein began to ponder its greatest limitation: the requirement that the laws of physics appear to be the same only for observers moving at a constant speed. In technical terms, motion at a constant velocity—that is, motion consistent with Newton's law of inertia—is *inertial motion*. One frustrating aspect of this limitation is that it is almost impossible to find an example of true inertial motion. Everything on the Earth, for instance, is experiencing some degree of acceleration all the time: the Earth spins on its axis, taking everything on its surface along for the ride, and the Earth moves in a near-circular orbit around the Sun. It was unsatisfying to Einstein that the principle of relativity would hold rigorously only for things in a state of motion that never really occurs.

In 1907, Einstein was still working at the patent office; his fame had not yet translated into a scientific career. One day, while he was pondering the problem of non-inertial motion, he had what he considered "the happiest thought of my life."

> Just as in the case where an electric field is produced by electromagnetic induction, the gravitational field similarly has only a relative existence. Thus, for an observer in free fall from the roof of a house there exists, during his fall, no gravitational field—at least not in his immediate vicinity. If the observer releases any objects, they will remain, relative to him, in a state of rest, or in a state of uniform motion, independent of

> their particular chemical and physical nature. The observer is therefore justified in considering his state as one of "rest."[2]

When we imagine an object falling under the influence of gravity, we tend to think of gravity as a force pulling on that body. What Einstein realized is that this picture is incorrect: a person or object—or cat—in freefall under gravity is *weightless*; it feels no force of gravity whatsoever.[3] Astronauts in orbit around the Earth are weightless because they are constantly falling toward the Earth—it just so happens that they are also moving parallel to the ground, so they are in a state of perpetually falling and missing the ground.

Almost immediately after having his happiest thought, Einstein laid the foundation for a new relativistic theory that incorporates gravity by introducing the *equivalence principle*. For our purposes, the equivalence principle may be summarized as follows:

> An accelerated motion is physically indistinguishable from being within a uniform gravitational field.

To understand the idea behind this, let's imagine a person in an enclosed rocketship with no windows. Like the people in Galileo's ship, the person has no way to see movement. If the ship is on the Earth's surface, the person will feel a gravitational force pulling downward. If the same rocketship is in space, away from any gravitational bodies, and is accelerating upward, the person will also feel a force pulling downward, closer to a gravitational body. The downward force is the inertial resistance of the body to acceleration. Einstein argued that there is no experiment that the person in the rocketship can do to determine which situation he or she is in—the two situations are physically equivalent. Anyone who has ridden an elevator has experienced this—when the elevator accelerates upward, a person

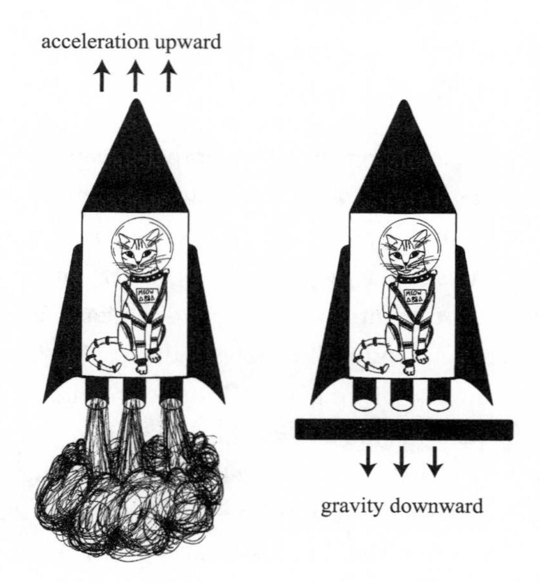

The equivalence principle. The force that a person in a rocketship accelerating upward will feel is physically indistinguishable from the force the same person will feel while standing still in a gravitational field pulling downward. Drawing by Sarah Addy.

in the elevator feels heavier. As the elevator decelerates near the top of its motion, the person feels lighter.

This was a profound insight into the nature of acceleration and gravity, but codifying it into a rigorous mathematical and physical theory took Einstein almost a decade of more work, along with significant assistance from mathematicians. At last, in November 1915, he presented the mathematical foundations of what would later be named the *general theory of relativity* in a meeting at the Prussian Academy of Sciences. Among the novel, even mind-boggling implications of the new theory was the idea that mass warps, or curves, both space and time. A massive body like the Earth or a black hole lies in the bottom of a "well" in spacetime. When close to a gravitational body, time moves slower, and a clock near the surface of the Earth will run slightly slower than a clock in an airplane.

In the context of this new theory, any object moving along the shortest path through spacetime—be it a straight or curved path—experiences no forces at all, is weightless, and may be considered to be in inertial motion. The weight we experience on the surface of the Earth may be reinterpreted as arising from the fact that we are blocked from following this shortest path, which would be a freefall toward the center of the Earth. Therefore Einstein's happiest thought is readily interpreted in terms of a universe of warped space and time.

This brings us, at long last, back to the question of falling cats. When a cat is dropped, it is in freefall and reflexively flips itself right-side up. However, according to Einstein's thinking, a cat in freefall will be completely weightless, experiencing no force in any direction—so how does it know which way to rotate in order to land upright? This question became a significant one for physiologists at the beginning of the twentieth century and would eventually bring them into contact with Einstein's profound theory.

The physiology work of Étienne-Jules Marey was primarily concerned with the motions an animal makes to achieve a particular goal. He was concerned with questions such as How does a cat move in order to turn over? and How does a bird flap its wings in order to achieve flight? Equally important and of interest to researchers, however, was the question of how the brain of an animal controls and coordinates the muscles of its body to produce these effects.

Neuroscience, the study of how the brain and nervous system function, had developed dramatically in the nineteenth century in parallel with, and in cooperation with, the physiological research of

Marey and others. Unfortunately for the creatures under investigation, Marey's anti-vivisectionist stance was not tenable for a lot of neuroscience research. At the time, the only way to test the function of various parts of the nervous system was to selectively damage those parts and examine the effect on the animals. This approach, sadly, would be continued in neuroscientists' study of the cat-righting behavior.

A cat can right itself in freefall in a fraction of a second; the speed of the reaction makes it clear that it is, at least in part, a reflex reaction. The term *reflex* refers to the involuntary reactions of living creatures to external stimuli; the common example is the patellar tendon reflex, tested when a doctor taps below the kneecap with a rubber mallet to elicit a knee jerk. By tracing the long history of researchers studying reflexes, we can see how it inevitably leads to the falling cat.

The study of reflexes has its origins in the work of René Descartes (1596–1650), who, we have seen, allegedly threw cats out of windows. If he ever performed such a test, his objective would have been to prove that animals are soulless machines that transform exterior stimuli into motor action; that is, he sought to demonstrate that the behavior of animals is simply a collection of automated responses. In his studies, Descartes championed the idea of *mind-body dualism*: that the mind is an entity separate from the material body and not subject to the physical laws that govern the body. In Descartes's view, the human mind (or soul) controlled the body through the pineal gland in the brain. Modern science has discounted the idea of dualism and treats the thinking processes of humans and animals alike as lying solely in the brain.

The term *reflex* itself can be traced to Thomas Willis (1621–1675), Oxford professor and founding member of what would eventually become the Royal Society of London. In 1664, Willis published an

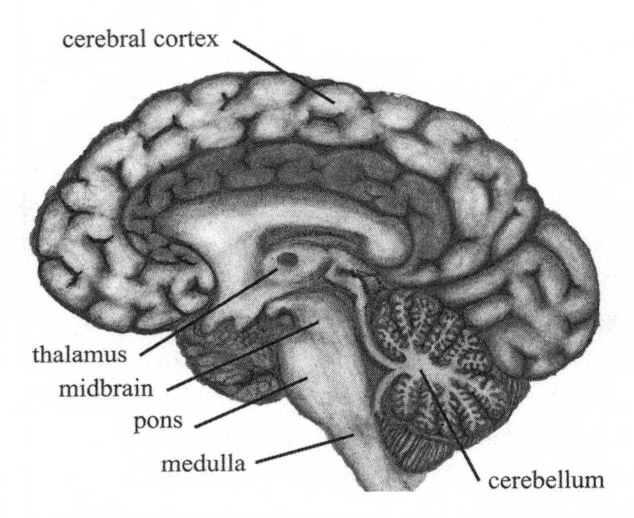

The main anatomical portions of the brain. Drawing by Sarah Addy.

important text on brain function titled *Cerebri anatome*. In it, he speculated that sensory inputs such as sights and sounds are directed to the cerebral cortex of the brain, resulting in conscious perception and memory. However, he imagined that some of those inputs were "reflected" back to the muscles through the cerebellum, resulting in automated movements, or "reflexes."

Willis, an English doctor, performed countless autopsies, which enabled him to elucidate much of the anatomy of the brain. It can be divided into three major components: the *cerebrum*, the *cerebellum*, and the *brainstem*. The exterior of the cerebrum is called the *cerebral cortex*, or simply the cortex, and it contains all the nerve cells, or neurons, involved with higher brain activity: the "gray matter" of the brain. The *pyramidal tracts* are nerve fibers that send messages from the cortex to the brainstem and the spinal cord, and the *thalamus* relays information from the sensory organs to the cortex. The cerebellum, which lies underneath and behind the cerebrum, is involved in the coordination of muscle movements and specific behaviors such as posture and balance. The brainstem, which may be further divided into midbrain, pons, and medulla, is responsible for automatic functions of the body, such as breathing and heart function.

Progress in reflex research was slow in the early days. Following Willis, researchers believed for over a century that the brain was the control center of reflex actions and that the spinal cord was only a conveyor of information—essentially a collection of wires—from the senses to the brain. This assumption was proven false by the Scottish physicist Robert Whytt (1714–1766), who showed in 1765 that a frog without a head can still reflexively react to external stimuli. Whytt's observations indicated that the spinal cord is the true mediator of reflexes. Whytt further showed that many reflex responses can be traced to particular segments of the spinal cord: different reflexes are controlled from different parts of the cord. In the immediate aftermath of Whytt's work, scientists concluded that conscious actions are controlled by the brain and that reflex actions are controlled by the spinal cord.

The reality, as we might imagine, is rather more complicated, and research in the nineteenth century led to major improvements and refinements in our understanding of reflex action. At the beginning of the century, the Scottish surgeon Charles Bell (1774–1842) demonstrated that there are two distinct types of nerves moving information through the body. He distinguished between sensory nerves, which convey sensory information to the central nervous system, and motor nerves, which convey commands from the central nervous system to muscles and organs. Bell published his insights in 1811 in a book titled *An Idea of a New Anatomy of the Brain*.

The English physiologist Marshall Hall (1790–1857) built on Bell's observations and formulated the first complete theory of reflex action, introducing the term *reflex arc* to describe the complete process by which a reflex is acted out.[4] Hall's theory, which is accurate in its broad details, can be explained by looking at the patellar tendon reflex. The doctor's mallet tap provides a *stimulus* to the knee tendon, which excites a sensory nerve, which transmits a signal to a segment of the spinal cord, which sends a command back through a motor nerve to the appropriate quadriceps muscle to cause it to jerk.

The knee-jerk reflex is also an excellent example of a crucial property of reflexes, also discovered by Charles Bell, known as *reciprocal inhibition*. When the patellar reflex is triggered, there is not only a signal from the spinal cord telling the quadriceps muscle to jerk but also a signal to the opposing hamstring muscle, an *antagonist muscle*, telling it to relax. Because of this inhibition of the antagonist muscle, the opposing muscles do not fight each other, which would not only be a waste of energy but could potentially cause muscle damage. As we will soon see, inhibition of muscle action plays a very large role in the nervous system. In an 1823 paper on the muscles that make the eyes move, Bell described such a response.

> The nerves have been considered so generally as instruments for stimulating the muscles, without thought of their acting in the opposite capacity, that some additional illustration may be necessary here. Through the nerves is established the connection between the muscles, not only that connection by which muscles combine to one effort, but also that relation between

> the classes of muscles by which the one relaxes while the other contracts.[5]

One other key discovery was made in Bell's era that illustrated the complexity of reflexes. Researchers found that a single reflex response can depend on multiple stimulus inputs that can change the strength and nature of the reflex, and that these inputs can include various senses as well as conscious brain control. A classic reflex of this form is the pain withdrawal reflex, which causes a hand to withdraw from a hot surface or an open flame. However, as action movie villains like to demonstrate, it is possible to resist this reflex and hold a hand in place over a fire even while in great pain.[6] The patellar tendon reflex, in contrast, is a simple reflex that cannot be controlled or resisted consciously.

By the late 1800s, then, there was a bewildering array of information on the functioning of the reflexes in particular and the nervous system in general, but no unifying theory connecting the disparate parts. In addition to reflex action, researchers had elucidated the anatomy of the brain, they had learned how different higher cognitive functions are localized in specific regions of the brain, and they had learned that the fundamental building block of the entire nervous system is the nerve cell, or neuron.

Into this state of affairs came Charles Scott Sherrington (1857–1952), an English physiologist and pathologist.[7] Sherrington, who would soon have a singular career, started life in a rather singular manner: records suggest that he was born about nine years *after* his official birth father, James Norton Sherrington, had passed away. His true father may have been a married surgeon named Caleb Rose, who had begun a relationship with Sherrington's widowed mother. Apparently to avoid scandal, Rose left the official paternity to the late James and placed himself in the official role of "visitor" in the Sherrington home, at least until his own wife passed away in 1880.[8]

Caleb Rose was encouraging to his official stepson, and Charles Sherrington received his initial impetus into medicine from him. Family financial troubles did not allow him to begin his studies at Cambridge, as he had hoped, but he distinguished himself as a student at the Ipswich Grammar School and, by 1875, had passed his preliminary examination in general education at the Royal College of Surgeons of England. By 1879 he was attending Cambridge as a non-collegiate student. He earned a membership in the Royal College of Surgeons in 1884 and a bachelor's degree in medicine and surgery in 1885.

Sherrington's impetus toward neuroscience came as a consequence of events at the Seventh International Medical Congress of 1881. At that meeting, a vigorous argument arose on the localization of functions in the brain, which immediately sparked several experiments to resolve the question. Sherrington was not in attendance at the meeting, but he was brought in to assist in the follow-up experiments; the experience left a significant impression on him.

In spite of his interest in neuroscience, Sherrington's early work alternated between physiology and pathology (the study of diseases and their causes). He made a number of trips in the 1880s to study cholera outbreaks on the European continent. In 1887, however, he took a job as lecturer in systematic physiology at St. Thomas' Hospital in London and there turned to studying neurophysiology as his primary occupation.

His initial research focused on the knee-jerk reflex, and in 1893, he published the results of his investigations.[9] In this work, Sherrington made the important discovery of what are now referred to as *proprioceptive reflexes*, which play a key role in reciprocal inhibition. It was already known, because of the work of Charles Bell, that reflexes involving antagonistic muscles undergo reciprocal inhibition. But Sherrington investigated these reflexes in more detail and found that the *degree* of inhibition depends on whether the antagonistic muscle is initially flexed or not. A flexed hamstring, for example, would receive a strong inhibition, while a relaxed hamstring would receive a weak inhibition. Evidently, the knee-jerk reflex was not only receiving a signal from the patellar tendon but also getting information about the current state of the hamstring from sensory nerves within it. These sensory nerves, which are embedded in muscles, joints, and tendons, became known as *proprioceptors*; they provide information to the central nervous system about the stresses and strains produced by a living body on itself.

Sherrington had therefore demonstrated that the communication lines along which reflexes operate are much more sophisticated than had been previously thought. We may imagine the early view and the updated view of reflex action as different responses to a fire alarm. In the first view, a fire alarm was pulled, causing the fire department to respond, but without detailed information about the emergency. Sherrington showed that reflexes act much more like a call to the fire department through 911, in which lots of information about the fire is relayed before the department decides upon the appropriate response.

In further studies of antagonistic muscles, Sherrington noted another peculiar phenomenon that would have tremendous implications for the understanding of the nervous system. In animals that had their cerebral hemispheres completely removed, the extensor muscles (such as the quadriceps) became rigidly extended, a phenomenon soon labeled "decerebrate rigidity." The rigidity suggested to Sherrington that muscles, even when at rest, are constantly being excited by one part of the central nervous system and being simultaneously inhibited by signals from the cerebral hemispheres. In turn, this suggested to him that inhibition plays a much greater role in the nervous system and reflex action than had previously been imagined.[10] Sherrington would, in 1932, be a co-winner of the Nobel Prize in Physiology or Medicine for this discovery and would describe the phenomenon in his Nobel lecture.

> The reflex therefore, which at first sight seems a purely excitatory reaction, proves on closer examination to be in fact a commingled excitation and inhibition. Usually clearly demonstrable in the simple spinal condition of the reflex, this complexity of character is yet more evident in the decerebrate condition.[11]

In 1906, Sherrington further cemented his reputation in neuroscience with the publication of his book *The Integrative Action of the Nervous System*. In it, he developed the first unified picture of how reflexes work, from the level of cells to the level of the brain.[12] He introduced the concept of a *synapse* as the key connection point between nerve cells and as the location where excitatory and inhibitory reflex signals interact to determine total response. Sherrington used his synaptic view of reflex action, combined with evolutionary theory, to explain how and why major brain structures like the cerebellum and cerebrum came into existence.

Sherrington's novel views of reflex and brain function instigated many physiologists and neuroscientists to test his plethora of ideas. One of these researchers was Lewis Weed of Harvard Medical School, who in 1914 published his "observations upon decerebrate rigidity."[13]

It had been shown by Sherrington that the cerebral hemispheres produce an inhibitory effect on the muscles and that the removal of those hemispheres results in rigidity. Weed, conversely, was interested in locating the part of the brain or spinal cord where the *excitatory* effect on the muscles originates. Through extensive testing on cat subjects, he concluded that two parts of the brain play a key role in maintaining rigidity: the cerebellum and the midbrain. The cerebellum receives the impulses from the limbs that request a rigidity response, but it also simultaneously serves as a link connecting the inhibitory signal from the cerebral cortex to the limbs. Weed found the midbrain to be the origin point of the signals that are sent to the limbs to cause rigidity.

Weed was motivated to perform his studies "not only because of the physiological interest attached" but also because decerebrate rigidity possesses a very close similarity to the rigidity that develops in a number of deadly human diseases, such as meningitis. A better understanding of the nature of decerebrate rigidity would, in Weed's view, help diagnose and treat diseases related to the nervous system.

Considering the wealth of novel concepts that Sherrington had introduced about reflex function, and the common availability of cat test subjects, it was inevitable that the falling cat reflex would come under the scrutiny of neuroscience. In 1916, Weed, now at Johns Hopkins University, joined with his colleague Henry Muller to perform the first study of the cat-righting reflex from the neuroscience perspective.[14]

Their inspiration and motivation was again decerebrate rigidity, plus a hypothesis put forth by Sherrington in his influential 1906 book.

> The muscles that it [decerebrate rigidity] predominantly affects are those which in that attitude antagonize gravity. In standing, walking, running, the limbs would sink under the body's weight but for contraction of the extensors of the hip, knee, ankle, shoulder, elbow; the head would hang but for the retractors of the neck; the tail and jaw would drop but for their elevator muscles. These muscles counteract a force (gravity) that continually threatens to upset the natural posture. The force acts continuously and the muscles exhibit continued action, tonus.[15]

Sherrington suggests that the reason that rigidity is always "on" and actively suppressed by the cortex is because those muscles are the ones that keep animals upright against gravity. From a survival perspective, this makes perfect sense: an animal's ability to hunt or escape predators depends on its ability to move about upright, so those muscles should be active all the time by default. Sherrington says that nerve pathways that meet this evolutionary need for anti-gravity reflexes must be the origin of decerebrate rigidity.

Since the cat-righting reflex is arguably an anti-gravity reflex, albeit of a very different nature, it seemed worthwhile to Muller and Weed to study that reflex not only to test Sherrington's hypothesis but also to illuminate how cat-righting works neurologically. No high-speed photographs were taken for their experiments; they were interested not in the specific motions a cat makes to turn over but in isolating the way the nervous system initiates those motions.

They found, perhaps unsurprisingly, that a decerebrate cat does not exhibit any righting reflex at all, suggesting that higher brain function, even consciousness, is necessary for cat-righting. It is therefore a complex reflex arc, more like the pain withdrawal reflex than the knee-jerk reflex.

More significantly, Muller and Weed undertook an investigation into what *senses* a cat uses to determine which way to turn to land right-side up. A key focus of their investigation, and those that would follow, was the *vestibular system*, which provides the sensation of acceleration for living creatures. Though its functional parts are primarily contained in the inner ear, it might be considered a sixth sense to go along with sight, hearing, touch, taste, and smell. The system can be further compartmentalized into two distinct components: the *semicircular canals*, which detect acceleration in the form of rotation, and the *otoliths*, which detect straight-line acceleration.

Each ear contains three fluid-filled semicircular canals perpendicular to one another. The three canals allow the sensing of three perpendicular rotational movements, namely pitch (falling forward), yaw (spinning around the axis of the spinal cord), and roll (falling sideways). Rotation of the head causes the fluid in the canals to flow, exciting small hairs that send signals to the brain, indicating movement. Adjacent to the semicircular canals are the otoliths, which also distinguish linear acceleration through the motion of hairs. The otoliths can be further divided into the utricle, which is oriented horizontally and detects side-to-side and forward-to-back acceleration, and the saccule, which is oriented vertically and detects up-and-down acceleration.

Muller and Weed were interested in determining the relative roles of the vestibular and the visual systems in the righting reflex of cats. Through experiments, they found that blindfolded cats can rotate and land accurately and that non-blindfolded cats with a damaged

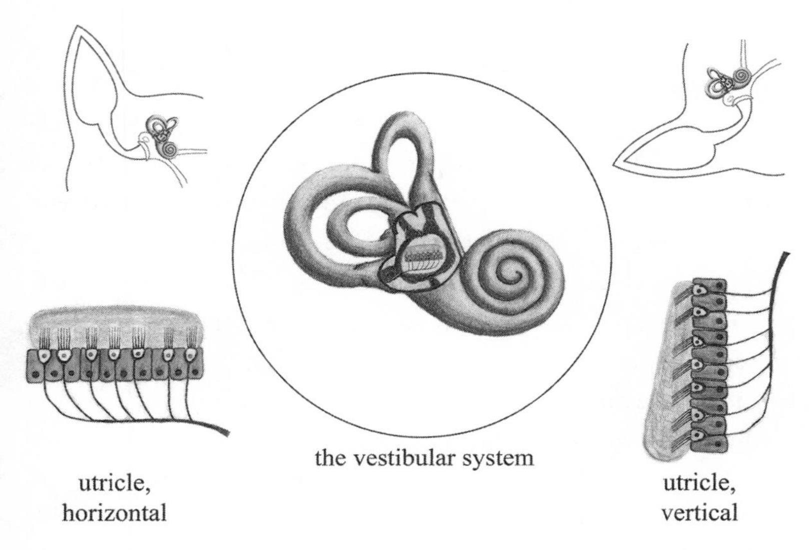

Simple illustration of the vestibular system, showing the semicircular canals (the three loops) and the utricle oriented horizontally and vertically. Drawing by Sarah Addy.

vestibular system can similarly land accurately. A cat that has a damaged vestibular system and is blindfolded would not make any effort to flip at all. These observations indicated that the righting reflex depends on either the eyes or the vestibular senses to determine the proper orientation for landing.

It is striking, in hindsight, that a blindfolded cat can flip over properly. As we have noted, according to Einstein's happiest thought, a falling cat feels no acceleration—the vestibular system is not active and the cat has no visual cue to show which way is down. How does the cat land in the correct position? Einstein's theory of general relativity was not on the minds of physiologists in the early 1900s, and the answer would come much later.

Though Muller and Weed made several important observations about the neurological basis of the cat reflex, they did not make any progress in confirming Sherrington's anti-gravity hypothesis. Their conclusion: "The results here recorded surely offer no evidence either for or against the hypothesis that the muscular reactions in decerebrate rigidity are the result of an attempted resistance of gravity."

The same question was being considered, about the same time, by the German researcher Rudolf Magnus (1873–1927).[16] Magnus, whose education in Heidelberg initially led him into a productive career in pharmacology, found himself drawn into a physiological study of reflex action after reading Sherrington's classic text. Magnus ended up meeting Sherrington at the Seventh International Congress of Physiologists in Heidelberg in 1907, and in 1908 he spent his Easter holiday with Sherrington to study how the body position of an animal affects its reflexes. In particular, Magnus was interested in understanding the role of reflex action in maintaining body posture; in Sherrington's terminology, Magnus wanted to understand how an animal maintains its anti-gravitational behavior while standing, bending, and walking.

The work would occupy Magnus for almost the next fifteen years, culminating in 1924 in a classic text on the subject, *Körperstellung*, and a lecture on animal posture before the Royal Society of London in 1925. To understand the reflexes that Magnus was interested in, we can do no better than review the list that he himself gave.

1. Reflex standing—In order to carry the weight of the body against the action of gravity, it is necessary that a certain set of muscles, the "standing muscles," should have by reflex action a certain degree of enduring tone, to prevent the body from falling down on the ground.

2. Normal distribution of tone—In the living animal not only do these muscles possess tone, but also the other muscles of the body, especially their antagonists, i.e., the flexors. Between these two sets of muscles a certain balance of tone exists, so that neither set of muscles gets too much or too little tone.
3. Attitude—The position of the different parts of the body must harmonise with each other; if one part of the body be displaced, the other parts also change in posture, so that different well-adapted attitudes, evoked by the first displacement, will result.
4. Righting function—If by its own active movements or by some outside force the body of an animal is brought out of the normal resting posture, then a series of reflexes are evoked, by which the normal position is reached.[17]

"Righting function" in Magnus's research applied initially to those reflexes that keep an animal standing, not to a falling cat righting reflex, though Magnus quickly related the two. He was aided in this research by Adriaan de Kleijn, a Dutch researcher who joined Magnus's lab as an assistant in 1912, and by G. G. J. Rademaker, a Dutch surgeon who entered into the collaboration much later, in 1922. A key result of their studies was the discovery of what are now called the *Magnus–de Kleijn neck reflexes*. These are a set of proprioceptive reflexes triggered by the turning of the head: the limbs on the side of the body toward which the head turns are tightened while those on the opposite side are relaxed. This reflex is most prominent in animals experiencing decerebrate rigidity and is considered part of the righting function: if the animal's head turns to one side, the animal adjusts its muscle tone to resist falling in that direction. This work was considered so significant in the science of reflexes that

Magnus and de Kleijn were strong contenders for a Nobel Prize in Physiology before Magnus's untimely death in 1927, at the age of fifty-three.

Magnus's studies on neck reflexes were largely done with cats, so it was natural to next investigate whether the same neck reflexes play a role in initiating the righting reflex of a cat in freefall. Magnus published his research on the topic in 1922 in a paper with the title "How the Falling Cat Turns Around in the Air."[18]

To test his hypothesis, Magnus needed his own high-speed photographs of a falling cat. Though he was aware of Marey's work, Magnus did not initially have access to Marey's images and ended up taking his own. He employed a camera system purchased from Heinrich Ernemann, an entrepreneur who had started producing motion picture cameras for amateurs in 1904; motion picture technology was already becoming commercialized. Magnus's published images are shown here.

In the paper Magnus describes the process by which the reflex acts.

> According to what has been said, the reaction in free fall is about the labyrinth of the head, by way of which the head is turned away from its normal position. This turn away is followed by the cervical reflex, by way of which the body follows the head, first with the thorax, then with the pelvis. In this way, there is an extremely fast-running helical movement of the animal through space which is introduced from the head.

In short, Magnus imagines that the acceleration of the "labyrinth" (another name for the region of the vestibular system) triggers the

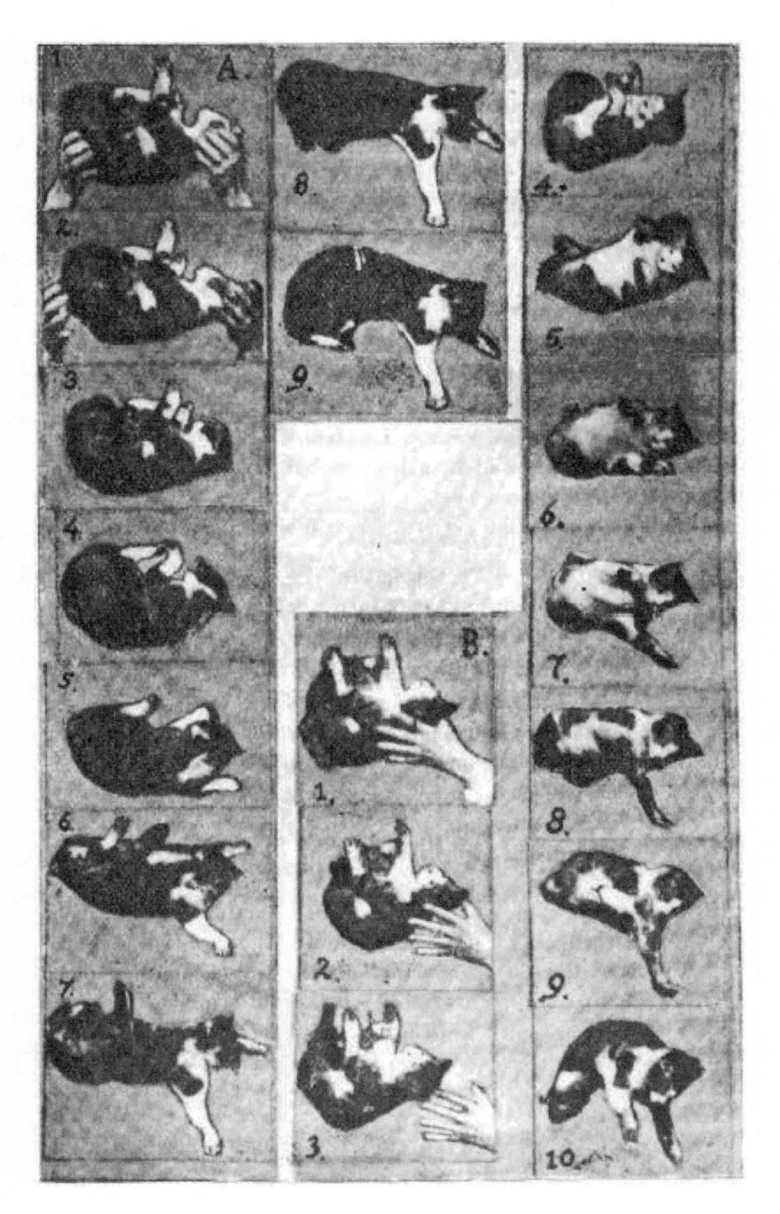

Rudolf Magnus's images of a falling cat, 1922. Hathi Trust Digital Library.

cat to begin turning its head. This head turn triggers the Magnus–de Kleijn neck reflex, which causes the rest of the body to follow suit in rotation, more or less making the whole cat twist like a corkscrew until it is right-side up.

Magnus was not a physicist, and we will see that his explanation of the reflex was in conflict with not one but two distinct physical principles: the equivalence principle and conservation of angular momentum. This would not be proven for another decade, but the result of the follow-up research would be the recognition of a previously unappreciated, and crucial, motion that a cat uses to right itself when falling.

The new research would be published in 1935, by Magnus's former lab assistant Gijsbertus Godefriedus Johannes Rademaker, along with his colleague J. W. G. ter Braak, working at the Physiological Laboratory of what is now Leiden University.[19] Rademaker had begun his career as a surgeon, getting his degree in 1912, and ended up spending five years in practice in Indonesia, starting in 1915. The intense workload and devastating effects of tropical diseases left their mark on him, and when he returned to the Netherlands, he opted for a career change. The physiology work in Magnus's lab was perfectly suited to him. There he studied the neurological aspects of muscle tone in cats and rabbits; in his doctoral work he elucidated the role of the *red nucleus*, a structure in the midbrain, in controlling posture.[20]

In spite of working closely with Magnus, Rademaker seems to have been troubled by his late employer's interpretation of the cat-righting reflex. His paper with ter Braak opens directly with a critique.

> The labyrinth-like reflexes take effect when the head, and with it the labyrinths, are not in "normal position." They are triggered by changes in the position of the labyrinths in relation to gravity. In these positional changes, gravity causes changes in the labyrinths that continue in the new position owing to the gravitational force. These changes trigger labyrinth reactions (the labyrinth reflexes) that return the head to the "normal position."
>
> In freefall, however, this influence of gravity ceases immediately. The animal thus turns its head in "normal position" during freefall, although the influence of gravity, which triggers the labyrinth reflexes, is canceled.

The authors here are indirectly referring to Einstein's equivalence principle. They argue that Magnus's explanation is inconsistent with this physical law. In freefall, according to general relativity, no force of gravity is experienced by the vestibular system at all; since the head-righting reflex is triggered by the head's being tilted *with respect to the force of gravity*, it is simply not possible for this reflex to be the deciding factor in the rotation of the cat.

Rademaker and ter Braak go even further to prove the point.

> The cat also turns in the air when the animal is thrown down quickly, so it is moved downward with greater initial acceleration than in freefall. Under these circumstances, at the beginning of the movement, the influence of gravity on the labyrinths is not only canceled but even replaced by a force in the opposite direction. Nevertheless, even then the cat turns, and the direction of rotation does not change.

One more argument shows that the head position alone cannot be a deciding factor in how or why a cat flips over: it is possible to hold a cat upside down, but with its head upright, as Rademaker and ter Braak showed. If the righting reflex were triggered by the cat's head falling in an upside-down position, then it should apparently not be triggered at all with the head upright. It was found that cats flip nevertheless.

The researchers did not discount the influence of the vestibular system altogether in this situation, but only argued that the falling cat reflex must be triggered in a fundamentally different way from the reflex that causes the head to return upright.

> From these observations, it can be seen that freefall righting cannot be due to the gravitational labyrinth reflexes. Turning around in the air is determined by the labyrinths. So there must be a second kind of labyrinth reflex that is not gravity-induced, but is based on excitement of the labyrinths as a result of the falling motion.

It is reasonable to think, then, that the *loss* of weight sensation in the labyrinths triggers the reflex. This does not explain, however, how a cat knows which way to turn even when blindfolded. Rademaker and ter Braak would not find an explanation.

Continuing their critique, Rademaker and ter Braak pointed out a second way in which Magnus's explanation was not consistent with known physics. They noted that the physical mechanism to which Magnus attributed the cat's rotation, a corkscrew twist of the cat from head to tail, violates conservation of angular momentum. In Magnus's model, all parts of the cat twist in the same direction in turn. But if the head turns right, the body must counterrotate to the

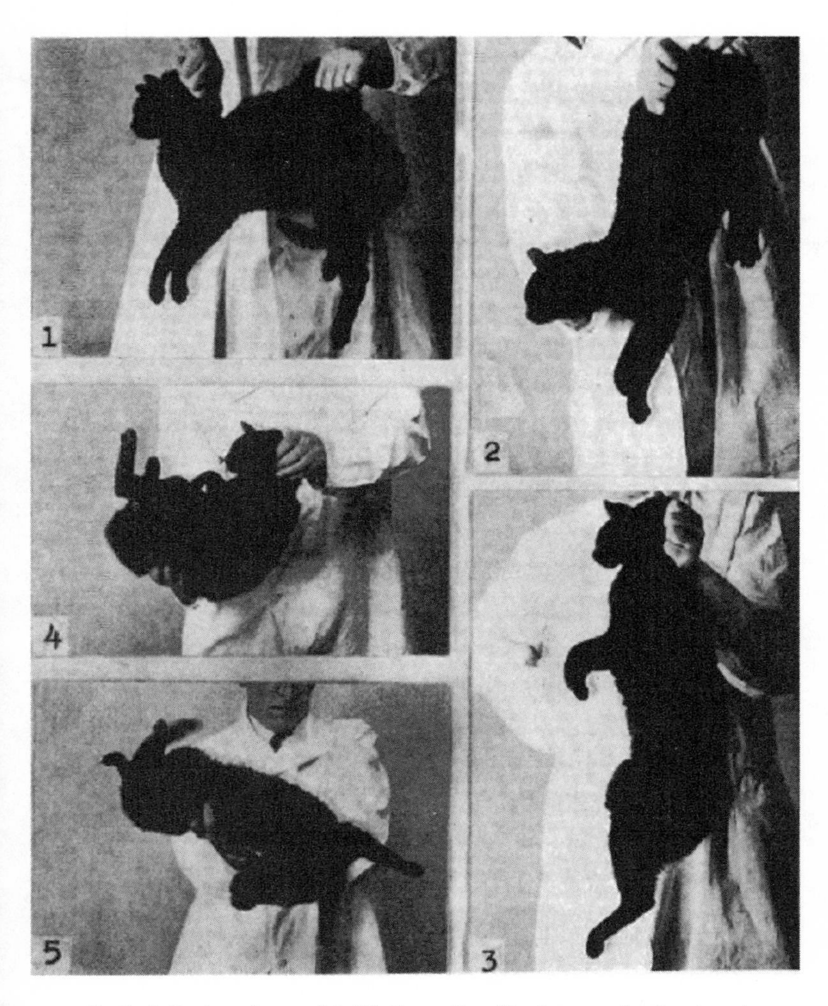

G. G. J. Rademaker and J. W. G. ter Braak's photographs showing the position of the head in falling cats. All cats, when dropped, land on their feet, even the ones in parts 4 and 5, which are upside down but with heads upright. From Rademaker and ter Braak, "Das Umdrehen der fallenden Katze in der Luft," Acta Oto-Laryngologica, *23:313–343, 1935, fig. 10. Copyright © Acta Oto-Laryngological AB (Ltd.), reprinted by permission of Taylor & Francis Ltd., http://www.tandfonline.com, on behalf of Acta Oto-Laryngologica AB (Ltd.), vi.9, http://www.informaworld.com.*

left to conserve angular momentum; if the body then turns right, the head must counterrotate to the left. In the end, the cat can gain no overall rotation in this manner.

The two Dutch authors also found the tuck-and-turn explanation of cat-turning given by Marey and Guyou to be unsatisfying. In order for a cat to flip completely over using this technique with only two tucks, it must rotate its head *more* than 180 degrees to counteract the counterrotation of the torso. This degree of rotation did not seem to be borne out by the photographic evidence.

Criticizing an existing hypothesis is arguably the easy part of science; coming up with a new hypothesis is much more difficult. Fortunately, Rademaker and ter Braak were up to the challenge. They proposed a mechanism for cat-turning that today is called the bend-and-twist model. They first noted that all models of a turning cat so far had assumed that the cat keeps its back straight during the motion, even though the photographic evidence clearly shows otherwise. Imagining a cat's body, for simplicity, as consisting of two cylinders that can bend and twist at the waist, they observed that the more a cat bends, the more the upper and lower body sections oppose each other. This idea is illustrated nearby. If the gray arrows represent the direction of angular momentum of the two parts, we can see that a fully bent cat will have the angular momenta of its upper and lower body sections cancel each other out. The cat is therefore able to rotate with an angular momentum of net zero.

The simplest way to see how this works is to imagine an upside-down cat, its body straight. Then it bends its body at the waist, making two parallel cylinders of its body, with the top of the head facing outward. Now it performs a 180-degree twist. Because the net angular momentum of the cat is zero, there is no overall change in the cat's orientation, but the top of the head now faces inward. When the

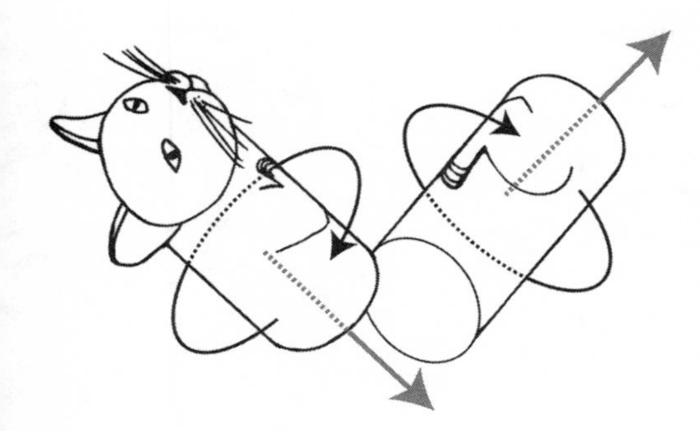

Rademaker and ter Braak's bend-and-twist model of cat-turning. Though both body sections are turning in the same direction when the cat is straight, they are turning in opposite directions when the cat is bent in half. Drawing by Sarah Addy.

cat straightens its back again, it is right-side up. No cat can twist with a fully bent body, but with a smaller bend, it can perform a longer twist to make up for any counterrotation it experiences.

This model is physically distinct from the tuck-and-turn model. In the tuck-and-turn, a cat changes the moment of inertia of its upper and lower body sections to allow one section to rotate more than the other counterrotates. In the bend-and-twist model, the cat pits the rotations of its upper and lower body sections against each other to allow a net change in orientation.

The authors provided mathematical results to back up their hypothesis, as well as an unintentionally hilarious "hot dog" illustration that shows how different muscle groups could contract to achieve the desired effect. This figure, included here, curiously suggests that the cat starts with its back arched outward, even though photographic evidence shows that it starts with its back bent inward, belly tucked.

In recent years, the bend-and-twist has become accepted as the most important part of a cat's rotation in the cat-righting reflex. Looking back on some of the earlier series of photographs, like Marey's side view of a falling cat in chapter 4, we readily see the bend-and-twist. In Marey's side view, the sixth image from the right

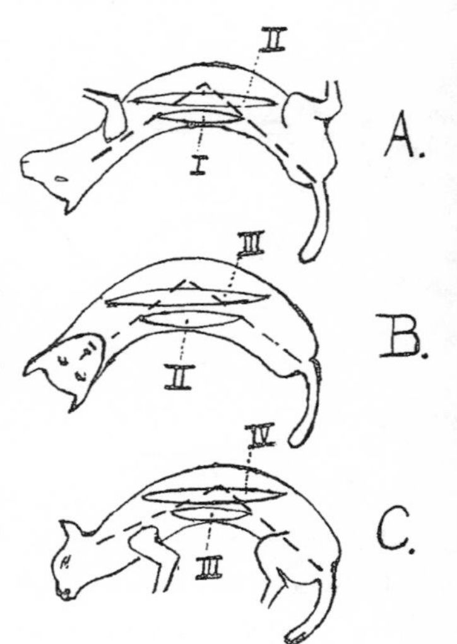

Rademaker and ter Braak's "hot dog with a bored demon face" illustration of the bend-and-twist method. From G. G. J. Rademaker and J. W. G. ter Braak, "Das Umdrehen der fallenden Katze in der Luft," Acta Oto-Laryngologica, *23:313–343, 1935, fig. 5. Copyright © Acta Oto-Laryngological AB (Ltd.), reprinted by permission of Taylor & Francis Ltd., http://www.tandfonline.com, on behalf of Acta Oto-Laryngologica AB (Ltd.), v1.9, http://www.informaworld.com.*

in the upper sequence rather clearly shows the cat midway through the motion, as in part B of Rademaker and ter Braak's "hot dog" illustration. Fredrickson's 1989 photos of a tailless cat (see chapter 6) exhibit it in the top row, in frames 3 and 4, counting from the left. The same motion can be seen in some of Rademaker and ter Braak's own photographs, such as frames 2 and 3 of figure 7 from their paper, included here.

Rademaker and ter Braak's paper was not the last word on how a cat moves its body to flip over in freefall, but it introduced what is almost certainly the predominant mechanism by which a cat achieves its extraordinary feat.

The persistent question remained, however: How does a cat know which way is up when it begins to fall? General relativity indicates that a cat experiences no force in freefall that the vestibular system can use for orientation, and nor is vision necessary, since blindfolded cats flip over without any trouble.

The bend-and-twist as shown in Rademaker and ter Braak's photographs long before it was given that name. From G. G. J. Rademaker and J. W. G. ter Braak, "Das Umdrehen der fallenden Katze in der Luft," Acta Oto-Laryngologica, *23:313–343, 1935, fig. 7. Copyright © Acta Oto-Laryngological AB (Ltd.), reprinted by permission of Taylor & Francis Ltd., http://www.tandfonline.com, on behalf of Acta Oto-Laryngologica AB (Ltd.), VI.9, http://www.informaworld.com.*

Giles Brindley playing the logical bassoon. From Brindley, "The Logical Bassoon," plate XIX. Courtesy of the Galpin Society.

The British physiologist Giles Brindley (b. 1926) hypothesized that the only remaining possibility is that the animal maintains a reflexive *memory* of which direction is down and uses that memory to instinctively fix the landing direction. In the 1960s he instituted a series of peculiar tests on rabbits, which also exhibit the falling reflex, to see whether his hypothesis was correct.

"Peculiar" could be said to be Brindley's scientific specialty.[21] In the 1960s, in addition to his physiological studies, he invented his own electronic instrument, the "logical bassoon."

In the rabbit tests, described in a pair of conference proceedings, Brindley set out to expose a rabbit to an acceleration that would, as far as the rabbit could tell, change the direction of gravity.[22] The rabbit would be subjected to the false direction of gravity and then released to see whether it fell true to the ground or true to the apparent direction of gravity.

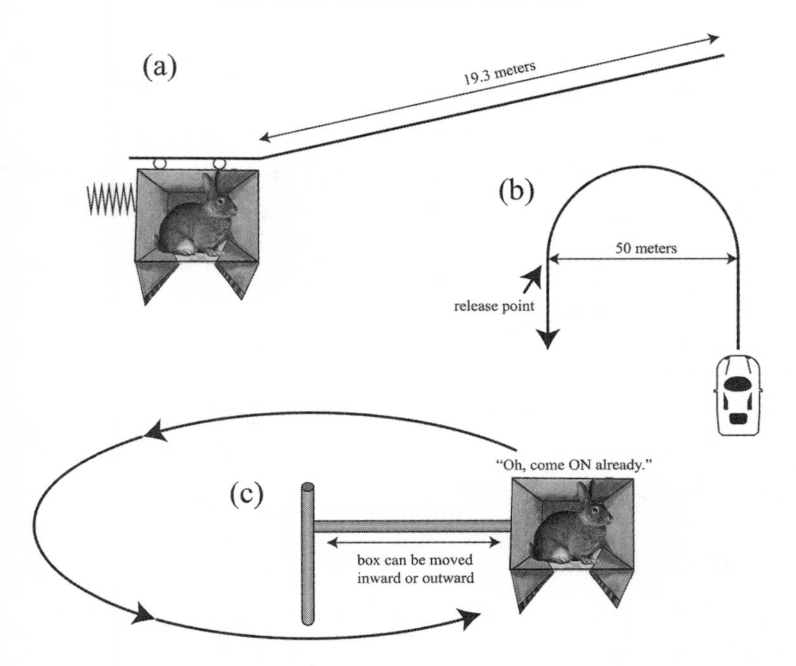

Brindley's experiments with a falling rabbit: (a) the "rabbipult"; (b) the "carabbit;" (c) the, uh, "centrabbitfuge"? Drawing by Sarah Addy and me, with my names and labels.

The conference proceedings do not provide a lot of detail about, or any photographs of, the apparatuses used for the experiments, so we must make some deductions about their specific forms.

The first experiment involved a rabbit in a box suspended from a pair of rails that inclined upward at 13°6′. The box was catapulted up and down the slope, and when it returned to the bottom, a trap door opened to release the rabbit, at which point its righting reflex in falling was photographed.

The catapult acceleration of the rabbit lasted 0.3 seconds; a full trip up and down the slope lasted 8.0 seconds. During that time, the box was tilted, and the rabbit felt gravity at a 13° angle. We must assume that the platform where the rabbit began and ended its ride was level. So if the rabbit experienced a tilted gravitational field for an extended period of time, but then the field was suddenly changed just before release, what happened to the rabbit? To quote Brindley:

> On the ninth second the floor of the box was automatically opened, and the rabbit was photographed while it was falling 150 cm (0.553 sec) to a cushion. It retained throughout the fall an inclination of about 13° to the upright. In contrast, if it was held stationary at 13°6′ and then dropped, it turned in the air and became upright within 30 cm.

In other words, the rabbit orients itself based not on what the gravitational field is at the very instant of falling but on what it was for some seconds before that. In hindsight, this makes sense: the moment an animal falls, it may experience any number of forces as it flails around and pushes/bounces off its perch. Memory of the direction of gravity in the recent past is more reliable than that memory at the instant it drops.

But Brindley wasn't done. He next took the rabbit in a "motor-car."

> The same box, without its wheels, was mounted in a car. A rabbit was put in as before. The car was driven at 32 km/hr in a straight line for 30 sec, and then suddenly turned without change of speed into a circular orbit of 50 m diameter, so that the rabbit suddenly experienced an apparent gravitational field inclined at 17°51′ to the vertical. After 10 sec the floor of the box was opened and the rabbit photographed while it was

> falling 80 cm (0.404 sec) to a cushion. It retained throughout the fall an inclination of about 18° to the upright.

On the circular path in the motor-car experiment, the rabbit experienced a centrifugal "force" that made the rabbit feel as though gravity had tilted toward the outside of the circle. As before, the rabbit fell according to the direction of the gravity it had historically experienced, not toward the vertical.

As was later recounted, this experiment was done on a disused runway at the Duxford Aerodrome. Brindley's wife, Hilary, did the driving while Giles did the photography.[23]

Both the railway and the motor-car experiments were discussed in Brindley's first conference paper; in the second, he recounted an additional test: spinning rabbits in a centrifuge.

> Rabbits were placed in a box with a spring trapdoor floor, mounted 105 cm from the axis of a centrifuge ("merry-go-round" of the Engineering Laboratories, Cambridge University). The centrifuge was brought to a speed giving here a gravitational-accelerational field inclined at about 30° to the vertical. After 1/2 min or more the box was moved quickly to the axis of the centrifuge, and at a time between 1/4 and 15 sec later the trapdoors were opened. . . . Those dropped 1 sec or less after the change usually fell in a very oblique posture roughly corresponding to the field at the periphery of the centrifuge. Intermediate times gave intermediate postures.

The takeaway from all of Brindley's experiments is that rabbits, and presumably cats, keep a "memory bank" of roughly the last six to eight seconds of the direction of gravity and fall based on what that memory bank tells them. To put it another way, after a dramatic change in the direction of gravity it usually takes about six to eight seconds for the animal to completely acclimate. Presumably, the semicircular canals of the vestibular system, which detect rotations, can keep track of how far the animal has rotated relative to that stored memory of gravity, even when it is blindfolded.

Though questions apparently remain about how, exactly, the central nervous systems of cats and rabbits manage the various sensory inputs and reflex actions in order to perform a reliable and accurate righting reflex, Brindley had adequately resolved how these animals manage to cope with Einstein's happiest thought. It took some forty-five years from Einstein's discovery for the mystery to be resolved.

It should be noted that in Brindley's experiments, the change in the apparent direction of gravity the rabbit experienced was increasingly extreme—from 13°6′ to 17°51′ to 30°. In his second conference paper, Brindley proposes an even more extreme experiment, in which a rabbit is taken along for a ride in a nose-diving airplane to give an apparent change in the direction of gravity of 40°. That experiment was apparently never carried out, probably because the U.S. Air Force was already carrying out similar experiments with cats in preparation for sending humans into space.

8
Cats . . . in . . . Space!

Around 1960, researchers at the Aerospace Medical Research Laboratories at Wright-Patterson Air Force Base in Dayton, Ohio, produced a movie to highlight the progress they had made in improving pilot safety and studying the effects of weightlessness on humans.[1] In one segment of the footage, men, cats, and pigeons are taken on board a modified C-131 cargo aircraft and subjected to weightless conditions for the entertainment of the audience. The cats, in particular, can be seen to right themselves just fine under normal gravity, but in the weightless state, they tumble about uncontrollably, not quite sure which way is up. The same thing happens to the pigeons; the men, presumably because they were prepared for the experience, fare much better.

This movie could be said to mark the culmination of early Air Force research into the effects of weightlessness on living creatures, research that had been going on for nearly a decade by that point. Cats not only played a key role in that research, owing to their natural righting talents, but would figure prominently in work that followed, as NASA tried to determine the best way for astronauts floating in space to change their orientation. Remarkably, cats would have a lot to teach humans as humanity took its first steps toward the stars.

The path to humans—and cats—in space began with a small group of idealistic rocket hobbyists in Germany in the 1920s.[2] These hobbyists, who dreamed of spaceflight, were high on enthusiasm but short on funds and supplies. Their struggles drew the attention of the German Army in the early 1930s, as it appeared they could provide the solution to a political problem. The Treaty of Versailles, signed after World War I, prohibited Germany from building up significant military forces or conventional military armaments. Rocketry was not included on the list simply because it had not been used as a weapon during the Great War. Given the loophole, Germany could develop a new long-range attack capability without drawing the wrath of the rest of Europe. After watching a rocket test of the amateur group, the German Army offered jobs to a number of the rocketeers, including the famous (and often infamous) Wernher von Braun. Thus began the research that would culminate in the horrific V-2 rocket attacks on London in World War II.

Von Braun and his colleagues were largely interested in space exploration, not weaponry, but the Nazi regime did not offer them much of a choice in the matter: they could join the Nazi party or die. They joined, and continued to develop rocketry until the end of the Second World War and the defeat of the Nazis. The Americans and the Soviets had not failed to notice the obvious potential of the new technology. In the chaotic aftermath of the war they rushed to recruit as many German rocket scientists as they could. The Soviets did their work mostly in one fell swoop, in Operation Osoaviakhim: on October 22, 1946, they "recruited," at gunpoint, some two thousand German scientists from Soviet-occupied Germany into their rocket program. The comparable American project, Operation Paperclip, occurred over a longer period, from 1945 through 1959. Initially, many of the scientists were housed, monitored, and debriefed in

Western-controlled Germany, but eventually many of them and their families emigrated to the United States to help with the space effort. Wernher von Braun and his colleagues, upon learning of the death of Hitler, immediately sought sanctuary with the American forces; von Braun had moved to the United States by the end of the year.

Among those recruited were the brothers Fritz and Heinz Haber, who emigrated to the United States in 1946. Fritz Haber, an aeronautical engineer, had worked for Junkers Aircraft in Germany during the war, designing a method by which a missile could be transported piggyback on an aircraft; a similar system would eventually be used to transport the American space shuttle on top of a modified 747. Heinz Haber, a physicist, worked as a Luftwaffe reconnaissance aviator during the war. Together, they were assigned to work for the U.S. Air Force School of Aviation Medicine, housed at Randolph Air Force Base in Texas; it eventually became part of the Department of Space Medicine. Curiously, they largely turned away from their respective fields of expertise and focused instead on physiology, in particular trying to understand the effects of weightlessness on the human body.

The School of Aviation Medicine had existed in some form since 1918, for the use of airplanes in the Great War led to a need to understand what sort of medical conditions pilots might experience. The Department of Space Medicine was formally created in 1949 to investigate what medical concerns might arise in space travel. The term *space medicine* was originally coined in 1947 by Hubertus Strughold, another German scientist brought over to the United States through Operation Paperclip. Strughold would become the first director of the Department of Space Medicine.

Though long-term stays in space are now common, in the late 1940s there was no way to predict what effect even a short bout with weightlessness might have on human physiology.[3] Gravity is omnipresent in our lives, and to those early researchers it was not clear how much our physiology depends on that constant force in order to function properly. As Heinz Haber wrote ominously in a 1951 magazine article,

> In most discussions of space travel the consequences for the passengers of this weightlessness have been taken lightly. In fact weightlessness evokes a pleasant picture—to float freely in space under no stress at all seems a comfortable and even profitable arrangement. But it will not be as carefree as it seems. Most probably nature will make us pay for the free ride.
>
> There is no experience on the Earth that can tell us what it will be like. True, the first instant of free fall in a dive from a diving board approximates the gravity-free state associated with ideal free fall, but it lasts only a moment.[4]

What sort of negative effects might humans expect in a weightless state? Heinz Haber, with his colleague Otto Gauer, expected that the respiratory and cardiovascular systems would be relatively unaffected. However, they worried that the same proprioceptive impulses that provide important information about the state and orientation of parts of the body might malfunction in extended weightlessness. A conflict between the information about orientation provided simultaneously by the visual and vestibular systems could lead to extreme disorientation or something like unending motion sickness. The proprioceptors in the muscles were also a concern. Since human bodies are, in effect, "calibrated" to function properly while experiencing a constant force of gravity, the loss of that gravity could throw that calibration off, making every movement by a space traveler hyper-exaggerated. If these hypotheses were confirmed, it would put severe limitations on humanity's future in space.[5]

Space was not the only concern, however. With the advent of jet-powered aircraft during World War II, planes were flying faster and higher than ever before, to altitudes where air resistance is a negligible factor in flight. Any aircraft gliding unpowered under such circumstances would effectively be in freefall, and the pilot of such a plane would be in a weightless state. So terrestrial concerns also motivated the researchers in space medicine.

The biggest difficulty in studying such effects was the absence of an extended weightless state on the Earth. As Haber had noted, a fall from a diving board gives a truly weightless state only for an instant. The same is true in skydiving: even in a jump from a hot air balloon, there are only a few seconds approximating true weightlessness before air resistance provides a sense of "down."

One option for creating an extended weightless state was the use of drop towers, which are elevators designed to fall freely from a height—and slow down before hitting the ground. But such towers can be built only so tall, giving passengers a weightless state for only a few seconds at best. The Haber brothers suggested a superior solution in 1950—the use of an airplane flying a parabolic trajectory.[6]

This strategy is illustrated in the nearby graph. Under the principles of general relativity, any object moving freely in a gravitational field is in a weightless state. This state can be approximated for passengers in an aircraft flying in an appropriate manner. The plane first accelerates into an upward trajectory, during which the passengers experience an increased gravitational force. Then, the airplane reduces its thrust and follows a parabolic trajectory, like a baseball tossed to a friend would follow. The pilot must keep just enough thrust to compensate for the air resistance drag on the aircraft, which

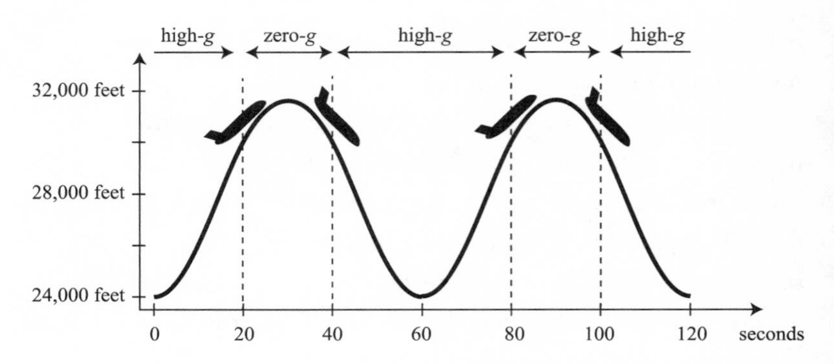

The roller-coaster parabolic trajectory for weightlessness. My drawing.

would otherwise result in an effective gravitational force. After the plane accelerates downward, inevitably the pilot must come out of the dive, subjecting the passengers to an increased gravitational force yet again. The process can be repeated if desired, making the overall aircraft trajectory a series of roller-coaster-like ups and downs.[7]

It did not take long for this scheme to be implemented, with some of the most daring test pilots making the first forays into extended weightlessness. In the summer of 1951 test pilot Scott Crossfield of Edwards Air Force Base produced zero-gravity states in both upright and inverted flight. Though he noted a sensation of "befuddlement" during the transition to weightlessness, he found, promisingly, that he had adapted to the sensation after his fifth flight. He also noted a tendency to overshoot while reaching for switches on his instrument panel, partly confirming Gauer and Haber's concern about malfunctioning proprioceptors.[8] Crossfield would become famous two years later as the first pilot to fly at twice the speed of sound.

Air Force test pilot Chuck Yeager, who made similar flights in 1952, also noted some disorientation, in particular a sensation of falling during the transition period, as well as some "orientational disturbances" during weightlessness, which disappeared once weight had been restored.[9] Yeager is famous for having been the first pilot to break the sound barrier in flight, which he achieved in 1947.

The first systematic study of the effects of weightlessness on humans was implemented at the Aero Medical Laboratory at Wright-Patterson Air Force Base, the place where Hubertus Strughold had moved in 1949 and which became another major center of space medicine.[10] In this study, a Lockheed F-80E Shooting Star fighter jet was modified to include a prone-position bed in the nose of the aircraft. The plane could be flown from either the bed or the conventional cockpit seat. In essence, a pilot could fly this aircraft lying down, though it seems that the standard practice was to have a test subject in the bed and a pilot controlling flight from the cockpit.

Typical flights involved eight to ten subgravity trajectories with an average duration of fifteen seconds. During the weightless periods, subjects were asked to accomplish a variety of coordination tasks, such as shaking their heads and reaching for objects. The participants responded very well, with minimal effects on orientation, and heart rate and electrocardiogram monitors showed no significant changes. The participants did feel that being restrained in a chair and having a visual reference to counteract any vestibular sensations helped them stay oriented. A free-floating blindfolded person might still suffer severe disorientation.

Weightlessness was not the only concern for researchers. In any proposed rocket flight into space, astronauts would be exposed to extreme forces, and these forces could impair—or even kill—an astronaut. Subjects in the fighter jet bed were therefore also subjected to extreme accelerations—measured in multiples of the force of gravity g, or "g-force"—and asked to describe their sensations. However, it was much easier to run high-acceleration tests on solid ground, using a rocket sled that could accelerate to extreme speeds and be decelerated suddenly. Such tests began in 1947 and continued into the 1950s. The most famous and frequent participant in these tests was Air Force Colonel John Stapp, who on December 10, 1954, subjected himself to a rocket sled ride that reached a peak speed of 632 miles per hour and exposed him to a stunning force of 46.2*g* in deceleration. This test simultaneously made Stapp the record holder for highest g-force intentionally experienced and for highest land speed, giving him the title of "fastest man on Earth." His work led to major improvements in the safety harnesses and seats in fighter jets. Somewhat remarkably, considering the extreme abuse he subjected his body to, Stapp lived to age eighty-nine, dying peacefully at home in 1999.

High-g tests were common in the 1950s, and consequently subgravity effects were the greatest unknown and concern for any future space travel. Animals were inevitably drafted into tests of such effects. Rockets were used for animal tests, as a sort of compromise between two ideal circumstances: duration and safety. Whereas passengers on airplanes were limited to intervals of about twenty seconds of weightlessness, a rocket could fly much higher and maintain a parabolic trajectory much longer, providing several minutes of weightlessness. Because rocket flight was new and appallingly risky, human tests were out of the question. Rocket testing with animals was the natural result.

For these tests, two types of rockets were used. The Air Force still employed the reliable German V-2 rockets of Wernher von Braun, but they were expensive to build. The military had contracted with the Aerojet Corporation in the late 1940s to build a cost-efficient alternative for research, the Aerobee. Five V-2s and three Aerobees were used in an initial study that involved launches from White Sands Proving Ground in New Mexico from 1948 to 1952. Monkeys and mice served as the test subjects in all launches. The monkeys were anesthetized and hooked up to monitors that could track their vital signs by radio through all stages of the flights; on some flights the mice were also hooked up to monitors, while on others the mice were filmed floating in a weightless state to see how they reacted.[11]

As it turned out, the concerns about safety were justified. All five V-2 rockets failed to deploy their parachutes, as did the first Aerobee. The second Aerobee landed safely, but a delay in returning to base with the animals after recovery resulted in the primate dying en route from heat exhaustion. Only on the third Aerobee flight were all animals safely recovered. But all the rockets used radio to transmit vital signs and had sturdy camera housings for the movie footage, so even the crashed rockets produced important data.

The primate studies confirmed what had already been seen in earlier work and hypothesized by Gauer and Haber: the cardiovascular and respiratory systems of the animals were unaffected by the weightless state. The movie footage of the mice showed that free-floating animals were somewhat disoriented, but those that had gained a foothold on a stable surface seemed undisturbed. This agreed with the observations of the human subjects in the zero-gravity flights: apparently having the support of a fixed surface or seat could reduce the confusion associated with weightlessness significantly. Mice with intact and damaged vestibular systems were used on the flights to compare their responses, with the damaged mice having been given time to adjust to life and coordination on the ground without their motion sense. It was found, notably, that the damaged mice seemed *more* comfortable in a weightless environment than the undamaged mice. The researchers speculated that the undamaged mice were caught off guard by the sudden change in their vestibular sensations, which caused confusion, while the damaged mice, sensing no change, could adapt quickly.

These observations were confirmed over the next year in a key series of experiments which, for once, were not done in the United States, but in Argentina. The researcher, Harald von Beckh, was born in Vienna, Austria, in 1917 into a family of physicians, and he continued in the family tradition, earning the title doctor of medicine in 1940.[12] He became a lecturer at the Academy of Aviation Medicine in Berlin in 1941 and also served as a pilot and flight surgeon. After the fall of the Nazi regime, von Beckh realized that it would be impossible to continue flight research in Germany for years; finding himself in Genoa, Italy, he volunteered at the Argentine consulate to continue his work out of Buenos Aires.

Von Beckh was interested in studying orientation and muscular coordination in a weightless environment, and South America provided him with the perfect animal test subject: the Argentine snake-necked turtle, *Hydromedusa tectifera*. Von Beckh outlined their ideal behavioral qualities.

> These turtles would seem to be especially suitable for studies of orientational behavior and muscular co-ordination, because of their ability to move under water with extraordinary speed and skill in all directions during their quest for food. The animals belong to an extremely voracious class of water turtles. Under normal gravity conditions, i.e., on the ground or in horizontal flight, they strike like snakes at their food, projecting their S-shaped necks with pin-point accuracy at the bait. They will also snatch a piece of meat hanging from the mouth of another animal. In fact, when they are hungry, they try to pull out the bait which is already in the mouths of other turtles.[13]

The Argentine snake-necked turtle. Photographed by Daiju Azuma in 2007, shared under CC BY-SA 2.5. © OpenCage. Wikimedia Commons.

So the Argentine turtles possess a natural ability to hunt in all directions, they strike with precision, and they are highly motivated, which made them desirable subjects for tests of coordination. One particular turtle provided another fortuitous quality for von Beckh. Having accidentally been left for multiple days in an overheated aquarium, the turtle had lost its vestibular function. At first, it struggled to strike accurately at food, but it gradually recovered its skills and after three weeks was able to eat normally. Von Beckh concluded that the animal had suffered a permanent vestibular injury but had adapted to using visual orientation to compensate. With this turtle and undamaged animals, von Beckh could make a good comparison of reactions to weightlessness.

The animals were taken into a two-seat jet in a cylindrical jar filled with water that was open at the top. During the zero-gravity dives, the turtles were fed samples of meat held in a pair of pincers, and their accuracy in striking was assessed. It is worth noting that zero-gravity flight is never perfect, and interesting effects can be had with open bodies of water in aircraft. Von Beckh noted that "in the transition from horizontal to vertical flight, some brief negative acceleration was produced. At this time the water (and occasionally the animals with it) would rise up, forming an ovoid cupula to a height of 20 or 30 cm, above the top of the jar. However, most of the water would flow back when the jar was lifted to the same height."

Von Beckh found that the undamaged turtles, like the mice in the earlier American rocket experiments, struggled to target the bait provided for them, while the damaged turtle performed as well as it had on the ground. As had been predicted earlier by Haber and Gauer, the intact animals were evidently disoriented by the conflicting information from their visual and vestibular systems; the damaged animal had no such conflict. The intact turtles, after twenty or thirty flights, gradually improved their hunting ability, which suggested that humans, after some initial struggles, could also adapt to a weightless environment and function normally.

Von Beckh also tested coordination in humans on zero-gravity flights. During weightless periods, subjects were asked to mark crosses in boxes on a sheet of paper. The subjects did quite well except in the weightless state with eyes closed; this result agrees with earlier observations that cats and humans alike use their vestibular system and their eyes in cooperation to coordinate movements. With both senses removed, the subjects were quite disoriented.

With so many studies of animals in a freefall environment, it is somewhat surprising that it was quite late in the research program

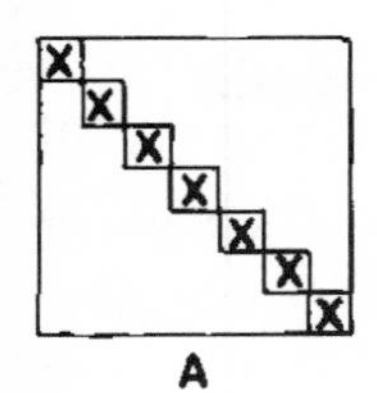

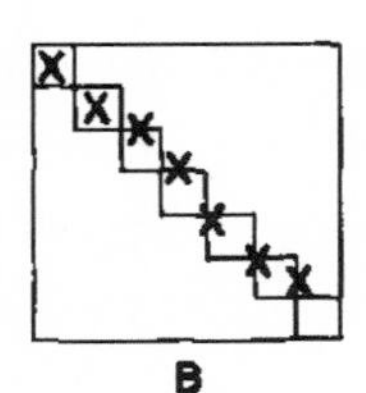

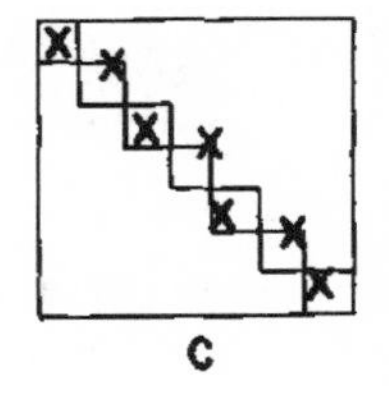

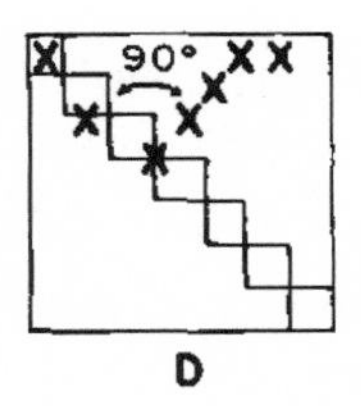

Cross-drawing test given in horizontal flight (a) with eyes open and (b) with eyes closed and in a weightless dive with (c) eyes open and (d) eyes closed. From Gerathewohl, "Subjects in the Gravity-Free State." Used with permission.

before cats joined the investigations. Finally, in 1957, Siegfried Gerathewohl, another German recruited by Operation Paperclip, and Major Herbert Stallings, a pilot at the Randolph School of Aviation Medicine, undertook a study of cats in zero-gravity flights. They described their motivation:

> From a practical viewpoint the question was asked how the righting reflex of the cat would work during sub-gravity and zero-gravity. Would the cat turn around when held upside down or would it stay in this position? Is there a time factor involved which may indicate adjustment and adaptation? How will other cues—for instance, visual orientation—affect the functioning of the reflex? The answers to these questions were sought not only to satisfy our own curiosity, but to clarify the role of the otolith organ during weightlessness.[14]

One troubling question about weightlessness remained unsettled: How would a weightless human fare when not secured to a seat or other fixed reference point? Researchers had not yet used large cargo planes for zero-gravity tests. The next best thing was to take small animals into smaller airplanes. Eight cats were taken on weightless rides: four that were three weeks old, two that were about eight weeks old, and two that were about twelve weeks old. Not all cats did equally well. Indeed, the youngest cats exhibited no righting reflex at all. The researchers concluded that cats develop this reflex between the fourth and sixth weeks of life.

For the experiments, the cats were held upside down and then released after different periods of weightlessness: one, five, ten, fifteen, twenty, and twenty-five seconds. Movie cameras recorded the reactions of the animals to zero-gravity. The aircraft available for the experiments were T-33 or F-94 training fighter jets, making for some surreal images of cats floating around in front of a pilot in an oxygen mask, and adding a new element of danger to being a fighter pilot.

The studies indicated that the cats were very good at turning right-side up when they had been weightless for at most five seconds. For longer periods of weightlessness, the success rate of turning dropped, and at a duration of fifteen to twenty seconds the cats failed as often as they succeeded. Blindfolded animals failed more often and earlier in the weightless state. These results appear to be in agreement with the studies of the "memory" of the rabbit-righting reflex that Brindley would do in the 1960s.

One key conclusion that Gerathewohl and Stallings arrived at in their work was a recognition that the otoliths—the parts of the vestibular system that detect linear changes in motion—are affected by changes in acceleration more than by acceleration itself. That is, constant acceleration does not affect the otoliths as much as the onset or removal of acceleration. Some additional experiments were done to see how a cat would respond to a negative g-force, one that pulled them toward the ceiling of the cockpit. Both of the adult cats turned in the apparent direction of gravity, one even standing inverted on the top of the cockpit canopy.

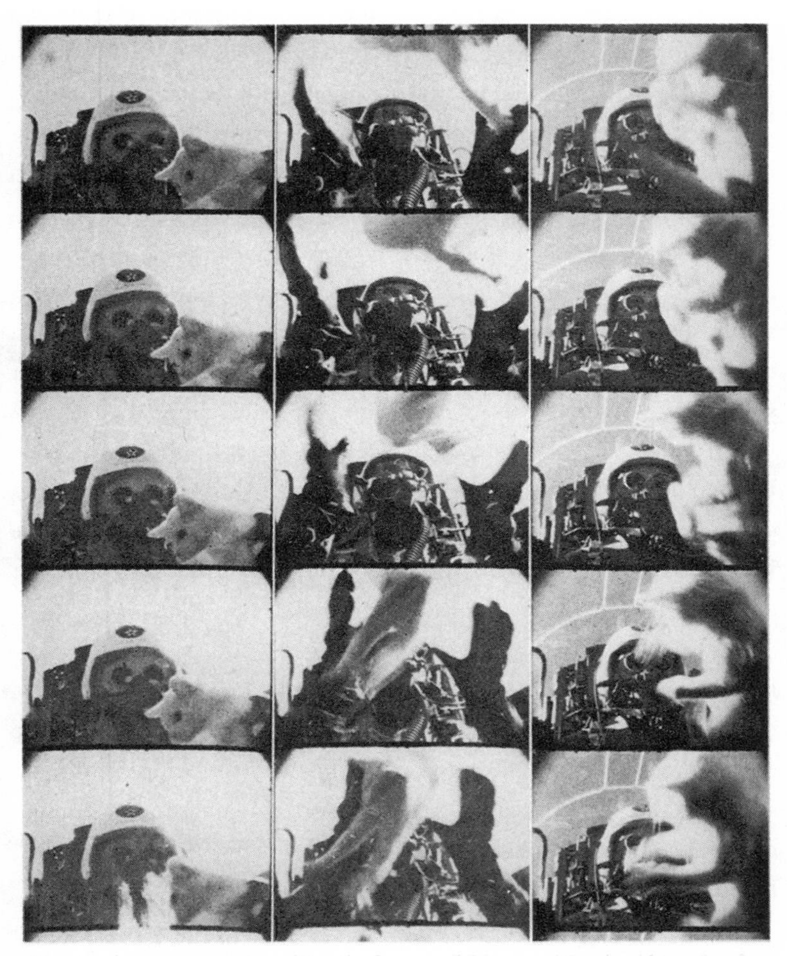

Fig. 1. Photograph of cat before the experiment in the T-33 aircraft.

Fig. 2. Prompt righting reflex of cat immediately upon entering the weightless condition.

Fig. 3. Delayed reflex: the cat turns slowly after a certain delay.

Three photo sequences of a cat being tested in a zero-gravity trajectory in a fighter jet. Major Stallings was the pilot. From Gerathewohl and Stallings, "The Labyrinthine Postural Reflex." Used with permission.

A similar study on cats and weightlessness was done several years later by Grover Schock at Holloman Air Force Base in New Mexico in 1961.[15] This study, which involved animals with damaged labyrinths as well as intact animals, confirmed earlier results that animals that have already adapted to having a damaged vestibular system seem to perform better in a weightless environment.

But the Gerathewohl and Stallings work of 1957 can be seen as the last major study of weightlessness in animals. That same year, two significant events happened that would change the nature of such research. The first was the introduction of a C-131B cargo plane for simulating weightlessness.[16] This plane, officially named the Weightless Wonder but soon dubbed the Vomit Comet by those who rode in it, could carry multiple passengers, untethered, for weightless rides of up to fifteen seconds in duration. It therefore became possible for humans to experience and study the weightless state without deducing its effects through intermediaries.

Using the C-131B removed most concerns and doubts as to how humans would react. As E. L. Brown described it, "Nearly everyone experiences a sense of exhilaration during zero g. It is a very enjoyable experience, and very relaxing. For the novice flyers, these enjoyable sensations are sometimes interrupted by extreme nausea, but because of the 2 1/2 g's immediately preceding and following the zero g period, it is impossible to say that zero g will cause nausea."[17]

A sense of that thrill can be seen in a contemporary photograph of such a ride, included here. In these early years, the movie footage at Wright-Patterson of cats flailing around in the Weightless Wonder was also filmed.

The second research-changing event was the launch of the satellite Sputnik 1 by the Soviet Union on October 1, 1957. This came as a shock to the Western world, and the United States military redoubled

"Miss Margaret Jackson, Physiology Branch, Aero Medical Laboratory, the first woman to be free-floating during zero gravity, demonstrates a characteristic pose." From Brown, "Research on Human Performance during Zero Gravity."

its efforts in the space race, particularly in crewed missions. Any ill effects of weightlessness, if they existed, would be dealt with along the way: Americans were going into space, and as quickly as possible. On July 29, 1958, President Eisenhower signed legislation establishing the National Aeronautics and Space Administration (NASA). Project Mercury was approved on October 7, 1958, with the goal of putting a human into orbit and returning him safely, ideally before the Soviet Union. The latter aspect of the mission failed when Yuri Gagarin did a single orbit of the Earth on April 12, 1961, but the United States was close behind. On May 5, 1961, astronaut Alan Shepard made the first American suborbital flight, and on February 20, 1962, John Glenn made three orbits around the Earth. The Space Race was well and truly on.

With the physiological studies of living creatures in space moving from the theoretical to the practical stage, many of the original initiators of the work moved on to other projects. Fritz Haber moved into private industry, joining the Avco Lycoming company in 1954, where he co-developed some of the first gas turbine engines. He was later promoted to vice president for European operations.

Fritz's brother Heinz had an even more surprising change in career. In the mid-1950s he became the chief scientific consultant to Walt Disney productions and worked on a number of popular science projects with them, including a famous 1957 episode of the Disneyland television series, called *Our Friend the Atom*, extolling the positive aspects of nuclear power. He went on to write a book based on the episode, as well as a number of other popular science books.

Hubertus Strughold became the chief scientist of NASA's Aerospace Medical Division in 1962, where he worked on the designs of the pressure suits and life-support systems used by the Gemini and Apollo astronauts. Because of his accomplishments he became known as the "Father of Space Medicine," but he was haunted throughout his life by the suspicion that he had participated in human experimentation at the Dachau concentration camp during World War II. Strughold perhaps summarized best those early years of research with the U.S. military in comments he made at the ten-year anniversary of Randolph Air Force Base's Department of Space Medicine: "Our work was not always taken seriously by outsiders. People heard of us, smiled and shook their heads. To them we were 'crackpots' and 'wild men.' It is perhaps fortunate that our beginning was small, and very very inexpensive."[18]

The role of cats in space research did not end with the first humans in space. Researchers recognized that maneuvering in a weightless environment presented its own challenges. How, for instance, could an astronaut floating freely change his orientation without any angular momentum? The cat-righting reflex provided the only well-studied example of such a maneuver, and it consequently received renewed attention.

Air Force researchers used the Vomit Comet as a place to test various maneuvering strategies, sharing the space with Project Mercury astronauts in training. A number of pure engineering solutions were evaluated, with mixed levels of success. Using magnetic shoes, for example, test subjects walked on the ceiling of the C-131B while in a weightless state. A lot of difficulty was encountered in adjusting the strength of the magnets to make walking easy. If the magnets were too weak, walkers could easily kick themselves off the metal surface; if the magnets were too strong, walkers would be effectively glued in place on the ceiling. For future motion outside space vehicles, a pneumatic propulsion unit was introduced, consisting of a tank of compressed air worn as a backpack, connected by a hose to a hand nozzle that could be pointed in any direction to provide thrust. The military personnel in flight suits with these contraptions look very much like Ghostbusters in the movie of that name. To prevent tumbling, researchers tried to make conservation of angular momentum work in their favor: test subjects carried spinning wheels as gyroscopes to hold their orientation steady. Just as a bicycle stays upright while the wheels continue to spin, these gyros would prevent astronauts from rotating uncontrollably.

But astronauts would also need to be able to control their motion in the absence of any special equipment. For this, the mechanism by which cats turn over, rather than the physiology of it, was of great interest. In the early 1960s researchers at the University of Dayton collaborated with a scientist at Wright-Patterson Air Force Base to determine a number of strategies for astronauts to use in reorienting themselves with body motions alone. In 1962 they published the results of their studies in a fascinating technical report titled *Weightless Man: Self-Rotation Techniques.*[19] In this document, the authors provide nine techniques for astronauts to use in changing their orientation, several for each axis of rotation. Z-axis rotations are rotations with the spine as a central axis—the astronaut spins like a figure skater; Y-axis rotations are in the same direction as front flips and back flips; and X-axis rotations are in the same direction as cartwheels.

The names of the techniques are whimsical and were chosen that way "so they can be referred to and understood quickly by space personnel." The full list of maneuvers is given here:

- (Z.1) Cat reflex
- (Z.2) Bend and twist
- (Z.3) Lasso
- (Z.4) Pinwheel
- (X.1) Signal flag
- (X.2) Reach and turn
- (X.3) Bend and twist
- (Y.1) Double pinwheel
- (Y.2) Touch the toes

The descriptions of these maneuvers paint interesting pictures of weightless astronauts. The "double pinwheel," for example, is described this way: "This continuous rotary maneuver has been demonstrated with success in zero-G flights. The procedure is quite simple. With the legs and feet tucked and the arms extended straight out parallel to the Y-axis, rotate the arms simultaneously in conical motions."

Of most interest for our purposes are the first two on the list. Indeed, that the very first is the "cat reflex" hints at the prominence the cat ability had in the minds of the resarchers. From the description, it is clear that the researchers considered this reflex to function in the way Marey originally described, as "tuck and turn," with the cat selectively altering the moments of inertia of different body sections. Here is how they describe it for humans.

> With the body straight and arms down, spread the legs to the sides, then twist the entire torso at the waist about the Z-axis to the right or the left. Holding this twist, extend the arms straight out to the sides, draw the legs together and then untwist by rotating the torso back to its original position. When the arms are lowered to the sides, the subject should have the exact configuration of the limbs with respect to the torso that he began with, but the body as a whole will have rotated a finite amount.

The second maneuver on the list, "bend and twist," is essentially the human version of the cat-turning technique observed by Rademaker and ter Braak in the 1930s. You can do this maneuver yourself. Starting from an upright standing position, bend your upper body to the side and raise your arms out to your sides. Then rotate your upper body forward, keeping it bent, until it is bent to the opposite side. Once you lower your arms and straighten out, you will have rotated a finite amount opposite to the direction in which you twisted your body. You can do this maneuver—slowly and carefully—even while standing on the floor, and if done right, you will feel the twist in your feet against the floor. The Rademaker and ter Braak technique became known as the bend-and-twist technique thanks to this Air Force work.

"Bend and twist" is listed as both a Z-axis maneuver and an X-axis maneuver, since it provides a little rotation along both axes. With slight modifications it can be used to emphasize one or the other. Whether any astronauts practiced these techniques or were given them as standard training is unclear. They may well have found it easier to develop their own movements organically while flying on the Weightless Wonder.

The flexibility of the human body allows much freedom for rotation, but it can also lead to problems. Consider an astronaut out in space, with a rocket backpack calibrated to send the astronaut straight forward. If that astronaut stretches one arm to the side, the center of mass of the astronaut will shift slightly to that side, and the astronaut will rotate as well as move forward. Before allowing astronauts to propel themselves in space outside a spacecraft, it was necessary to understand how changes in the body position of an astronaut could change motion and stability. To this end, detailed mathematical models of humans were developed in which every body section was treated as a cylinder, sphere, ellipsoid, or brick. A discussion of these models resulted in one of the most romantic descriptions of the human body ever put onto the page:

> The human body is a complex system of elastic masses whose relative positions change as the appendages are moved.[20]

The results of the weightless maneuvering studies were originally going to be put to the test in space on the Gemini 9A mission on June 5, 1966, by astronaut Eugene Cernan wearing an Astronaut

Maneuvering Unit rocket pack. But because Cernan overexerted himself in preparation for the walk, his helmet visor fogged up, and the test had to be cancelled. The first untethered extravehicular activity (EVA) would not be accomplished until much later, on February 7, 1984, when astronaut Bruce McCandless II used the more sophisticated Manned Maneuvering Unit (MMU). The MMU, looking very much like a high-tech easychair, possessed twenty-four thrusters that could be activated from the pilot's armrest to adjust rotation, orientation, and thrust.

The United States was not the only country studying weightless maneuvering. The Soviet Union had its own bioastronautics program in the 1960s. It extended over multiple facilities, complete with zero-gravity aircraft, centrifuges, and underwater training to simulate low-g environments. Like their American counterparts, the authors of a 1965 report gave due deference to the falling cat problem in their introduction, albeit with the history a little mistaken.

> Many experts in mechanics previously considered that a living creature cannot turn his body about some axis in an unsupported position. As their basic argument they cited the law of conservation of moment of momentum (law of areas).
>
> . . .
>
> The erroneousness of such assertions was proved by Deprez. He took several photographs of a falling cat which, without particular difficulty, always turned feet downward. It seemed that this fact was inexplicable from the point of view of the fundamental law of mechanics, namely, the law of areas.[21]

It was Marey, not Deprez, who took the photographs, and it was Lévy who first convinced the French Academy in 1894 that cat-turning was physically possible. Deprez was, in fact, Marey's strongest opponent at first.

To test their own strategies of self-rotation on the ground, the Soviets used a "Zhukovskiy bench," a horizontal platform containing a free-rolling sphere. A person standing on the platform could test strategies for horizontal rotation—for example, by swinging one arm above the head in a conical movement, which would cause the body to rotate in the opposite direction (the lasso, in Air Force parlance). To test more general maneuvers, the Soviets had subjects attempt them while bouncing on trampolines. The astronauts were trained in these maneuvers in order for the movements to "become automatic such as those of gymnasts, acrobats, divers, and other athletes who also must perform complex turns in the suspended phase."

Most of the simple maneuvers considered by the Soviets and the Americans are slow in execution. The lasso will eventually result in an astronaut facing in the opposite direction, but only after many circles of the arm and, consequently, many seconds of time. Cats can turn over in a fraction of a second. NASA was very interested in learning whether humans could flip over just as fast as felines. To find out would require much more sophisticated models and much more rigorous mathematical methods.

It so happened that a researcher had already, in the 1960s, been working on similar problems. Thomas R. Kane, professor of engineering mathematics at Stanford University, had developed a mathematical formalism that allowed the motion of complex systems of interconnected masses in a weightless environment to be analyzed. Space researchers had already recognized the usefulness of artificial gravity for astronauts making extended stays in space; one possible way to create artificial gravity is to use a spinning spacecraft or station; the centrifugal force would provide an outward force indistinguishable from a gravitational one. In 1967, Kane and his colleague T. R. Robe studied the stability of a satellite consisting of a pair of solid objects connected by some sort of partially elastic bridge, with the whole construction spinning around its center.[22] The usual falling cat model, consisting of a pair of cylinders connected by a flexible joint, is very similar.

Kane also attacked the problem of the motion of an astronaut in a weightless environment. Using his new mathematical technique, he showed how to computationally find optimal ways for the astronaut to change orientation.[23] This work drew NASA's interest—they awarded him a sixty-thousand-dollar grant to study such problems. Around the same time, Kane evidently came across the problem of the falling cat and, unsatisfied with the explanations given so far, took to evaluating it with his mathematical methods. What resulted is the most detailed and probably the most accurate mathematical model for the falling cat to date.

Kane largely agreed with the explanation given by Rademaker and ter Braak for the falling cat—namely, the bend-and-twist—but noted one major limitation. In Rademaker and ter Braak's model, the cat keeps the same bend angle between its upper and lower halves, which implies that it would land on its feet with its back in an inverse arch, exactly the opposite of what is actually seen. Kane, and his student M. P. Scher, argued instead that the cat begins with the Rademaker and ter Braak motion but gradually unbends its back as it twists until it is effectively facing sideways with an almost straight back. Then the cat bends to the other side, initiates another bend-and-twist motion, and completes its fall by landing with a bent back and extended legs. In short, in the Kane model the cat does the Rademaker and ter Braak motion twice, with a gradual change in back bend throughout.

A simpler way to visualize Kane and Scher's model is as three distinct motions in sequence. Imagine that the cat, as it begins to fall, bends and twists until it is facing sideways, bent right at the waist. Then the cat bends to the other side, until it is bent left at the waist. From this point, it can continue the bend-and-twist motion until it is bent forward at the waist and facing the ground.

In a paper published in 1969, Kane and Scher superimposed simulated solutions of the falling cat model on top of photographs of an actual falling cat.[24] The results were convincing. As Rademaker and ter Braak did, Kane and Scher modeled the cat as a pair of cylinders. For their model, they imposed an additional constraint: that the cat could not twist its upper and lower body sections relative to one another, as in the tuck-and-turn model of Marey.

This new work with cats was intended ultimately to help astronauts turn in a weightless environment; any methods devised would need to be tested by humans. As the Soviets had done, Kane used a trampoline as an inexpensive way to produce a brief weightless environment. To engineer a new turning technique Kane first developed the strategy for optimal cat-turning via mathematical equations. Then he input the motions into a computer that allowed him to sketch them in an understandable way. Finally, a skilled trampolinist, wearing a spacesuit, tested whether a human could reproduce the maneuver effectively.

The quirkiness of the project captured the attention of journalists at *Life Magazine* in 1968. The photographs taken for an article about Kane's work constitute one of the most surreal set of scientific images ever produced.[25] In the sequence, the images of a falling cat are shown side by side with images of a fully suited trampolinist mimicking the moves.

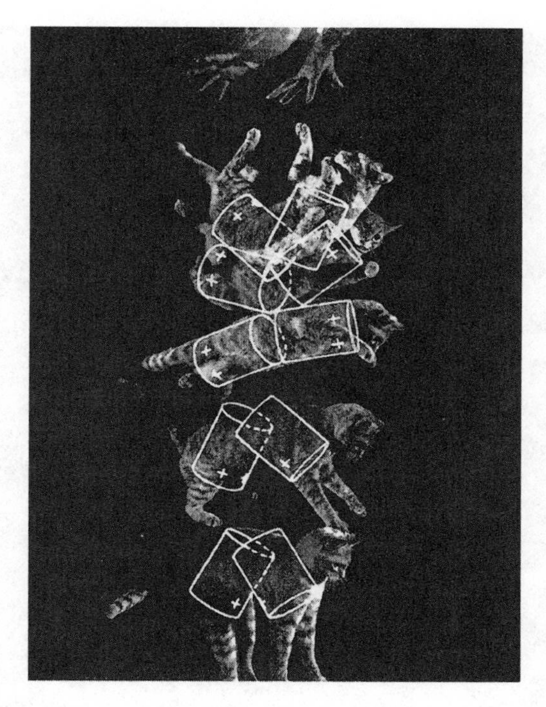

Computer-simulated Kane-Scher model of a falling cat superimposed on photographs. Reprinted from T. R. Kane and M. P. Scher, "A Dynamical Explanation of the Falling Cat Phenomenon," International Journal of Solids and Structures, *5:663–670, copyright 1969, with permission from Elsevier.*

Kane and Scher's publication seems to be the last one relating to falling cats and astronaut maneuvers. The two scientists followed up with an official publication detailing their self-rotational strategies specifically tailored for humans in 1970.[26] Falling cats would be of interest to researchers in other projects, but their role in space exploration came to an end in 1969.

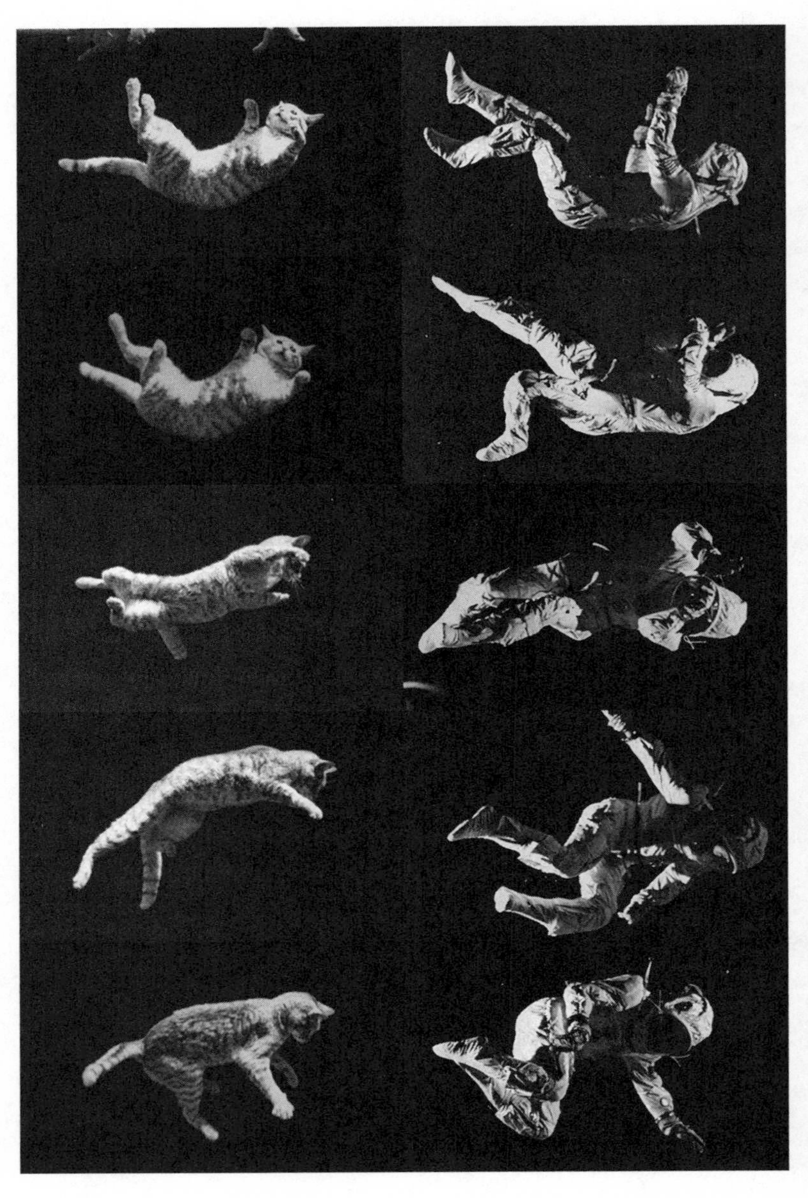

Falling cat and a trampolinist in a space suit. Ralph Crane/The LIFE Picture Collection/Getty Images.

In spite of all this research, only one cat has ever successfully made a trip to space and back safely. In the early 1960s, at the height of the early space race, France was also actively investigating the physiological effects of high forces and weightlessness on living creatures. Félicette, a stray cat found on the streets of Paris by a pet dealer, was purchased by the French government to become one of fourteen cats in their space testing. All the cats had permanent electrodes affixed to their brains to measure neural responses. On October 18, 1963, Félicette became the first cat launched into space, reaching a height of 156 kilometers (96.9 miles) on a thirteen-minute flight that included five minutes of weightlessness. She returned safely to Earth under parachute in her capsule.

Sadly, three months after surviving the dangerous ordeal, Félicette was euthanized so that French scientists could investigate whether the voyage had caused any physiological changes. In spite of her key role in the French space program, she was largely forgotten for many years. Then, in 2017, Matthew Serge Guy launched a

A postcard of Félicette, the space cat, printed after her successful journey. Translated, the inscription reads, "Thank you for your participation in my success of October 18, 1963."

crowdfunding campaign to erect a statue in her home city of Paris to commemorate her flight. The money was raised, and, as of this writing, the organizers are searching for an appropriate site for the monument, a permanent memorial to and reminder of the sacrifices that animals have made on our behalf.

Kane and Scher used a newly available technology to simulate complicated systems: computers. With sophisticated computer models available to describe the motion of humans and cats alike, it was only a matter of time before someone attempted to make machines that could act out their motions in real life. Before that would happen, however, cats revealed a new mystery in the way they fall.

9
Cats as Keepers of Mysteries

Though we have so far seen cats used primarily as laboratory subjects, they have also served as companions and even laboratory assistants to many scientists. Not every collaboration between feline and human is a productive one, however. In 1825, for example, several British newspapers reported a cat-related disaster.

> The celebrated Manheim telescope, the master piece of the famous Spaiger, an Hungarian optician, was destroyed a few days ago, in a most singular manner. A servant of the Observatory having taken out the glasses to clean them, put them in again, without observing that a cat had crept into the tube. At night, the animal being alarmed at the strong power of lunar rays, endeavoured to escape; but the effort threw down the instrument, which, falling to the ground from the top of a tower, was broken to pieces.[1]

The American poet Anthony Bleecker assumed that the cat was killed in the fall, and was inspired to write "Jungfrau Spaiger's Apostrophe to Her Cat," an imagined lament by the astronomer's daughter on the loss of her beloved pet.[2] Here's a small sample.

> Oh, tell me, puss, 'tis what I dread the most,
> Did some Kilkenny cat make thee a ghost?

Canst thou not speak? Ah, then I'll seek the cause.
What see I here? the bloody prints of paws;
And, oh, chaste stars! what broken limbs appear!
Here lie thy legs; the telescope's lie here.
The telescope o'erturned,—too plain I see
The cause, the cause of thy cat-astrophe.

The story is tragic for both cat and astronomer—assuming that it actually happened. There are good reasons to doubt the tale as reported, however. For one thing, the director of the German Mannheim Observatory—two n's—between 1816 and 1846 was Friedrich Bernhard Gottfried Nicolai; there is no record of a Spaiger to be found. It is also difficult to imagine an expensive, state-of-the-art telescope being mounted precariously on the roof of a tower where it could easily be toppled over. Nor is there mention of the calamitous event in any of the astronomy journals of the day. Work at the Mannheim Observatory seems to have gone on without hitch during the period in question.

We may also challenge the idea that Spaiger's cat, assuming it ever existed, died in the fall. The cat-righting reflex helps felines to survive unexpected tumbles, but the appearance of high-rise buildings in cities allowed cats to show off a complementary, perhaps even more baffling skill: cats not only have a high rate of survival when they fall from high places but they have a *higher* rate of survival from the highest falls. Not long after NASA lost interest in the cat-turning ability, veterinarians stumbled across this new puzzle.

New technologies, and corresponding changes in the way we live, often come with unexpected consequences. In 1887, for instance, the following complaint appeared in the journal *Science*. A. G. Thompson of Washington, D.C., wrote:

> Some disadvantage or evil appears to be attendant upon every invention, and the electric light is not an exception in this respect. In this city they have been placed in positions with a view of illuminating the buildings, notably the treasury, and a fine and striking effect is produced. At the same time, a species of spider has discovered that game is plentiful in their vicinity, and that he can ply his craft both day and night. In consequence, their webs are so thick and numerous that portions of the architectural ornamentation are no longer visible, and when torn down by the wind, or when they fall from decay, the refuse gives a dingy and dirty appearance to every thing it comes in contact with. Not only this, but these adventurers take possession of the portion of the ceiling of any room which receives the illumination.[3]

Those designing illumination for the nation's capital evidently did not reckon that they were also creating an ideal feeding ground for spiders.

Around the same time, in 1885, the world's first modern skyscraper went up in Chicago. Though other tall buildings had been erected for decades, the ten-story Home Insurance Building was the first to use structural steel in its frame. This innovation in material strength led the way to increasingly taller buildings, both commercial and residential—and to cats taking up residence at higher and higher altitudes. Inevitably, cats began falling from those heights. The builders of the Home Insurance Building certainly did not anticipate that their work would result in a new medical condition for felines, *high-rise syndrome*.

The phenomenon was identified and named by Gordon W. Robinson, the head of the Department of Surgery at the Henry

Bergh Memorial Hospital of the ASPCA in New York City, after he noticed an increased frequency of falling incidents.[4] Robinson's paper was published in 1976; it took nearly one hundred years after the first skyscraper was built for the syndrome to be recognized. Part of the delay, as Robinson noted, may be due to many pet parents not realizing that a fall had even occurred.

> The history may be confusing. Owners often do not see the accident occur and assume that the landlord, superintendent of the building, serviceman or a friend has entered the apartment whereby the cat escaped into the hall, down the stairs or fire escape and into the street or back yard where some type of trauma or poisoning occurred.

Robinson's paper, the first of its kind, was intended to draw attention to the syndrome and to assist veterinarians in recognizing it. He identified a triad of injuries that typically come with high-rise syndrome: epistaxis, split hard palate, and pneumothorax. *Epistaxis* is the technical term for a nosebleed, and *pneumothorax* is the term for a collapsed lung. The hard palate is the bony plate at the roof of the mouth that separates it from the nasal passages; in a high fall, a cat's hard palate will often split down the middle, front to back. In addition to these three primary injuries, teeth and other bones can be broken in a cat's fall.

Robinson nevertheless noted that cats can survive falls from astonishing heights.

> The distance cats have fallen and survived is nothing short of amazing. Our record heights for survival are as follows: 18 stories onto a hard surface (concrete, asphalt, dirt, car roof),

> 20 stories onto shrubbery and 28 stories onto a canopy or awning. Undoubtedly, these figures will bring forth "Letters to the Editor" whereby these heights have been surpassed with survival.

A little physics knowledge, however, reveals that it should not be quite so surprising to see that cats survive high falls, at least compared to humans. First of all, unlike humans, cats possess the righting reflex and typically land head up, which is crucial for survival. The relative smallness of cats also plays an important role. The adage "It's not the fall that kills you but the sudden stop at the end" is true: injury from a fall comes from the sudden *nonuniform* deceleration of a living creature's body. If an animal lands feet first, for example, the feet stop moving immediately, but the body above does not; the lower parts of the body are therefore subjected to the crushing inertia of the upper parts coming down on top of them. The more mass a creature has, the more injury its own weight does to it; cats have an advantage over humans in simply weighing less. Furthermore, a lightweight cat has a terminal, or maximum, falling velocity—at which gravity and air resistance are balanced—of about sixty miles per hour, roughly half the terminal velocity of a falling human.

Robinson's paper identified the problem of high-rise syndrome but did not provide any quantitative data on the survivability of cats or its relationship to the height fallen. Cats evolved to live and hunt and hide in trees and have adapted to falls from them. We might expect that cats are equipped to handle a one-story fall with negligible injuries, that falls from heights above this would be outside their regular experience, and that the number of injuries would increase with height, at least until the cat reaches terminal velocity.

The first comprehensive study to investigate such questions was performed in 1987 by Dr. Wayne Whitney and Dr. Cheryl Mehlhaff of the Department of Surgery of the Animal Medical Center of New York City.[5] They analyzed 132 cases of high-rise syndrome that came through their center over five months in 1984 and found that a remarkable 90 percent of cats survived the experience. Furthermore, they confirmed that on average the number of injuries increased as the height of the falls increased, but only for falls from eight or fewer floors up. Strangely, for higher falls, from above eight floors off the ground, the average number of injuries *decreased* dramatically, especially bone fractures.

The key graph from their paper is shown here. The injuries are divided roughly as Robinson originally characterized them, but with "fractures" replacing "nosebleeds." We can see that all types of injuries decrease significantly for cats that have fallen from nine floors high or higher. Amazingly, cats falling from extreme heights are typically *less* injured than those falling from moderate heights.

The results made national news, appearing in print countless times over the next few years. The *Los Angeles Times* presented Whitney and Mehlhaff's work in an article titled "They Land on Little Cat Feet." Two years later, the *New York Times* featured the work in an article with the painfully rhymed title "On Landing Like a Cat: It Is a Fact."[6]

The results are intriguing and counterintuitive—but are they accurate? Because nobody—thankfully—is actively throwing cats off roofs in controlled scientific experiments, studies on high-rise syndrome must rely on those cases that make it to the veterinary office, leaving open the possibility that the data have been biased in unexpected ways. For instance, suppose that, for the highest falls, cats with a large number of injuries die immediately. Those dead cats would

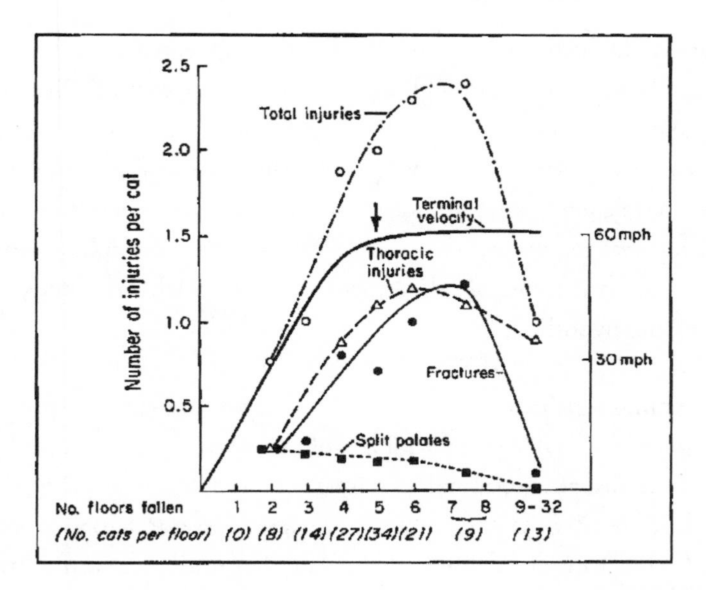

Graph showing injuries to falling cats from differing heights. From Whitney and Mehlhaff, "High-Rise Syndrome in Cats," p. 1399, reprinted with permission from the American Veterinary Medical Association.

not be brought into the vet's office, and the data of Whitney and Mehlhaff would be misleadingly skewed toward healthier cats with fewer injuries. The strange results may well have other causes, but this possibility cannot be discounted outright.

No follow-up study was performed for quite some time. Though cats regularly fall from high-rises, few places have buildings tall enough for high-floor data to be readily verified. Two more recent studies on high-rise syndrome, in Greece and Israel, looked only at cats falling from the eighth floor or lower.[7]

It took until 2004 for collaborating Croatian veterinarians in the city of Zagreb to collect enough data to test the Whitney and Mehlhaff results on extreme falls. In a study of 119 cases, they found that cats falling from the higher floors indeed appeared to have a lower rate of bone fractures, though thoracic injuries appeared to increase.[8]

Assuming that the data are not seriously misleading, we ask the next natural question: Why would cats falling from above the eighth floor have a lower rate of injuries, at least for some types of injuries? Noting that the decrease roughly coincides with the height at which cats first reach terminal velocity, Whitney and Mehlhaff offered the following hypothesis:

> As might be expected, the rate of injury in our cats was proportional to distance and speed of fall up to about 7 floors, a point just after terminal velocity is achieved. It was surprising, however, that the fracture rate decreased in cats falling > 7 floors. To explain this, we speculate that until a cat achieves terminal velocity it experiences acceleration and reflexively extends its limbs, making them more prone to injury. After terminal velocity has been reached, however, and the vestibular system is no longer stimulated by acceleration, the cat might relax and orient its limbs more horizontally, much like a flying squirrel. This horizontal position allows the impact to be more evenly distributed throughout the body.

This explanation is the currently accepted one, though its description of the physics is a bit misleading. In true gravitational freefall, a cat accelerates but does not "experience acceleration"—it is completely weightless. The sensation of true weightlessness would no doubt feel uncomfortable to a falling cat, causing it to extend its limbs downward. After the cat reached terminal velocity, it would feel its normal weight and might then have the sense to relax and stretch out its limbs horizontally for impact.

There is possibly more to the cat's strategy than relaxation, however. When the cat relaxes, its back may arch, making the cat cup air underneath its belly, creating a crude parachuting effect that could reduce its terminal velocity; as Whitney and Mehlhaff suggested, the cat might act "like a flying squirrel." This may be more than a crude analogy. In 2012 a Boston cat named Sugar survived a nineteen-story fall with only minor injuries.[9] Curiously, observations suggest that the cat may have used the flaps of skin under its arms to control its landing location, the way a flying squirrel uses its winglike membranes (patagia) to glide. Sugar landed in a pile of mulch surrounded by brick and concrete on all sides, which either indicates incredible luck or represents a small amount of flight control on the part of the cat. Nothing about what we might call cat gliding can be concluded from a single anecdote, but the thought is intriguing.

In any case, the average survival rate for cats suffering from high-rise syndrome is impressive. Most papers agree with Whitney and Mehlhaff's conclusion that about 90 percent of cats survive the ordeal. And the heights of fall are impressive. In the paper by Whitney and Mehlhaff, the record holder was a cat named Sabrina who fell thirty-two stories onto concrete, suffering only from a mild pneumothorax and a chipped tooth. In 2015 a Hong Kong cat named Jommi survived a twenty-six-story fall with no injuries at all, though Jommi had the good fortune of falling on—and punching through—a tent on the ground. Her owner recalls the event and Jommi's nonchalant reaction.

> We had left a little window crack open to get some fresh air and I suddenly had the horrific thought that she may have gotten through the gap.

> I looked over and saw a large hole in a tent 26 stories below, and I knew she had gone over.
>
> The force of her impact was so great it had bent the tent's aluminium frame so you can imagine my utter shock when I went inside and found her licking her paws as if nothing had happened.[10]

Cats may use a parachute-like strategy to survive falls, but some people have taken the comparison between cats and parachutists more literally than others. On February 15, 1967, the Parachute Association of Toronto presented a Free Fall Award to the cat Jasper,

Jasper being presented an award for his skydiving heroics. Image courtesy of York University Libraries, Clara Thomas Archives and Special Collections, Toronto Telegram fonds, ASC10068.

ward of Helen Kupari, for his "spectacular and historic free fall of fourteen storeys." Their proud photograph is included here.

One cat parent found a kindred spirit in her feline, who had a propensity for falling from high places. On January 26, 1972, a passenger DC-9 airplane exploded over Czechoslovakia, killing all on board except for the twenty-three-year-old flight attendant, Vesna Vulovic, who fell 33,330 feet to the ground—and survived.[11] She was in a coma for twenty-seven days and hospitalized for sixteen months, but in her rehabilitation she drew inspiration from her beloved cat Cicka, who had twice fallen out of a second-story window and suffered serious injuries, but had recovered both times.

Vulovic was eager to return to work as a flight attendant after the disaster, but the airline gave her a desk job, perhaps fearing that her presence on flights would be regarded as an ill omen and result in bad publicity. She passed away on December 23, 2016, and was mourned by the people of the former Yugoslav federation. She remains the Guinness world record holder for "highest fall without a parachute." The April 1973 issue of the magazine *RTV Revija* showed Vulovic and Cicka on the cover, with the title "Lady Luck's Favorites."

Studies on the science of high-rise syndrome continue to this day. In 2016 a group of Czech researchers and students suggested an alternative view of the cat's reaction to falls, suggesting that it is not the acceleration that causes the cat to reflexively flex its body but the *change* in acceleration.[12] This hypothesis is in agreement with the Air Force research of Gerathewohl and Stallings back in the 1950s. In a charming and simple experiment, the Czech researchers attached an accelerometer to a cat plush toy and dropped it out of a high-rise from floors at increasing heights. They found that the change of acceleration, which they dubbed the "coefficient of the cat's fear," is maximum at about the seventh floor, again roughly corresponding to the height at which injuries of falling cats begin to decrease.

Most research on high-rise syndrome, however, is focused on the medical aspects of the problem. There have been other studies on the injuries cats suffer, and at what heights, including research focusing on specific types of injuries and their treatment. The slips of cats past and present are being used to improve care and treatment of the falling cats of the future.

There is one more question to ask about high-rise syndrome: Why are so many cats falling out of buildings? Mehlhaff provided one explanation: "It has to do with coordination. We always give cats the reputation that they are supremely coordinated, which they are. But if you've ever watched two cats play and be silly, they can roll over and fall off whatever they are on. Sometimes, it just so happens that that is a ledge 21 stories up."[13]

Exceedingly skilled falling is not the only physics-based trick that cats have long kept secret. Even the most mundane cat habits have recently been found to offer scientific surprises. When MIT professor Roman Stocker was watching his cat Cutta Cutta drink from a bowl, he became curious as to how a cat laps up water. Humans can drink in a variety of ways—for example, by using suction to sip through a straw or by tipping water from a glass into the mouth. Animals like dogs, however, must drink by curling their tongue into a ladle-like shape to scoop the water up.

Cats seemed to be doing something different, but much too fast to see with the unaided eye. Stocker enlisted the help of MIT colleagues Pedro Reis, Sunghwan Jung, and Jeffrey Aristoff to study cat drinking with that most venerable of cat physics tools: high-speed photography. They started by patiently waiting around to get footage of Cutta Cutta drinking but moved on to filming other domestic cats

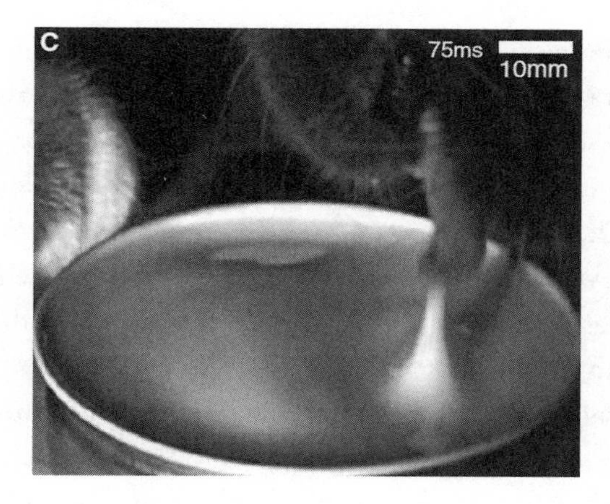

A cat lapping, showing the suspended column of milk before it is bitten off by the cat. From Reis et al., "How Cats Lap," fig. 1(c), reprinted with permission from the American Association for the Advancement of Science.

and, eventually, lions, ocelots, a tiger, and a jaguar. They bolstered their results with footage of other cat species found on YouTube.

What they observed was a previously unrecognized and remarkable strategy that cats use to drink. For all species of cats studied, the animal's tongue just barely touches the surface of the water before drawing back rapidly. Part of the fluid sticks to the cat's tongue, and the rapid withdrawal causes a thin column of water to be pulled upward into the air; the cat bites off the suspended column before it falls back into the dish.

Cats drink by taking advantage of intermolecular forces in the liquid itself, which cause a volume of liquid to be dragged along by the tip of the tongue. There is a perfect balance of inertial forces in the fluid and gravity, and through simulations the researchers were able to demonstrate that cats lap water at rates that maximize water intake. As in the problem of the righting reflex, it turns out that evolution had solved a problem in physics long before any human researcher recognized its existence.

The research has been verified, at least tongue-in-cheek, by another researcher. When asked to comment on the results, Steven Vogel of Duke University said, "Now that I've been clued in, I can report that what these people describe and explain agrees entirely with my own casual observations of the lapping action of the feline in charge of this establishment."[14]

After all of their work, the MIT research team learned that evidence of cats' remarkable drinking ability had been available for anyone to see for decades, thanks to the pioneering photography of Harold Edgerton (1903–1990). Edgerton did his doctoral work at MIT, where he studied the use of periodic high-speed electronic flashes with a device called a stroboscope in order to see the motion of fast-moving objects such as spinning fans. Realizing that the digital flash could be used to take photographs faster than previously imagined, he launched into photographing everything from bullets piercing apples, to athletes in motion, to the blasts of atomic bombs, to what was probably not the Loch Ness Monster.[15] During World War II he was commissioned to use his flash technology to do night aerial photography over occupied Europe, and he even flew on a number of dangerous missions to carry out the work.[16] In the 1950s he collaborated with famed oceanographer Jacques-Yves Cousteau.

In 1940, Edgerton was invited to Hollywood to demonstrate his techniques. The result of the ensuing collaboration was the 1940 film *Quicker 'n a Wink*. Included in the film is brief footage of a cat lapping up milk. The narration suggests that the cat curls its tongue down, making an upside-down ladle, but today we can clearly recognize the inertial effect presented in 2010. Appropriately, Edgerton did much of his high-speed photography work at MIT, where Stocker's group would make their discovery nearly seventy years later. In fact, the group used equipment from the MIT Edgerton Center, where, according to its website, "Harold 'Doc' Edgerton's spirit of discovery lives on" and "where we give students the opportunity to learn by doing."

Edgerton made his own footage of falling cats. In the 1930s, while he and his MIT partner Kenneth Germeshausen were still actively promoting their techniques to the scientific community, Edgerton found that the cat-righting reflex was a perfect attention-getter. Their cat images appeared in the *Science News-Letter* in 1934.

> No discredit to you or to any small boy for not finding out how the cat turns over. It required the resources of a great engineering laboratory, and the cleverness of two ingenious and hard-working young scientists, to make the matter plain. But a short time ago a movie of a cat turning over in midair, of a couple of flies "taking off," of a canary launched into flight and a number of other too-quick-to-see movements done by living animals were shown before the meeting of the National Academy of Sciences at Cambridge, and the most learned men in America ceased for a while their discussion of cosmic rays, the expanding universe, and other things of like abstruseness to watch and applaud.[17]

One of Edgerton's falling cat images is shown here. We can see part of the Rademaker and ter Braak motion in the cat's turn.

Cat tongues hold even more surprises. All cat parents are familiar with the feeling of being sandpapered when a cat licks them. That roughness of a cat's tongue has a very important practical use. One day, while the doctoral student Alexis Noel was watching her cat

Photographic sequence of a falling cat taken by Harold Edgerton. Copyright © 2010 MIT. Courtesy of MIT Museum.

Murphy lick a microfiber blanket, he became stuck but freed himself by pushing his tongue further into the blanket. Wondering why the tongue got stuck in the first place, Noel acquired a tissue sample of cat tongue and made a three-dimensional image of it using an X-ray computed tomography scanner. In reviewing the CT scan, she discovered that the tongue is not like sandpaper at all, but has a series of flexible clawlike spines that can twist around tangles in a cat's fur. As she describes it,

> When the tongue glides over fur, the hooks are able to lock onto tangles and snags. As the snags pull on the hook, the hook rotates, slowly teasing the knot apart. Much like claws, the front of the spine is curved and hook-like. So when it

> encounters a tangle, it is able to maintain contact, unlike a standard hairbrush bristle, which would bend and let the tangle slide off the top.[18]

Noel presented her results to the 2016 American Physical Society Division of Fluid Dynamics Meeting, and included high-speed photography of her cat grooming that shows how it flexes and twists its tongue to help it work out any tangles it encounters.[19]

Noel found one more surprise in the cat's tongue. When she exposed a sample of the hooked surface to a liquid, she found that the hooks are hollow and pull liquid into them via capillary action. This mechanism, she postulates, allows a cat to deliver saliva deep into its fur to help in the cleaning and detangling process. Noel has a patent pending for a hairbrush that would work on the same principle.

The investigations into cat drinking and cat grooming may seem to be lighthearted sidetracks, but in both cases the researchers have suggested that their studies give insights into how to build novel types of flexible robots. In fact, the cat-righting reflex became a subject of significant interest for robotics researchers and a sort of ultimate challenge for robot maneuverability.

10

Rise of the Robotic Cats

In late July 1994 scientists from Carnegie Mellon University traveled to Alaska with a bold mission: to venture into the crater of an active volcano and collect data. But the researchers, working on behalf of NASA, would not be going in themselves; they would send in Dante II, an eight-legged autonomous robot, to collect samples of toxic gases and map the topography of the crater's interior with lasers. The robot, connected by a tether to its control station on the crater's rim, would creep down to the bottom over the course of several days, making a return trip over a similar timeframe.

For the most part, the mission was a success. Dante II made it to the bottom, collected its data, and began the return journey. However, the weather had changed over the course of the mission, and hard snowy ground had melted into treacherous mud. On the way back up, Dante II slipped on a thirty-degree slope and toppled over, leaving it stuck. It took days to retrieve the roughly one-ton robot; a first attempt to lift it from its resting place by helicopter failed, forcing geologists to perform an in-person rescue operation, which arguably weakened the purpose, at least symbolically, of sending a robot into such a dangerous environment. The scientists attached a tether to the robot, allowing it to be lifted from the volcanic hotspot and, later, to enjoy retirement as a museum piece, with seven broken legs and a broken laser scanner.[1]

The robot's stumble was not unanticipated. It had been planned for in every way possible, and dreaded. As John Bares, a Carnegie Mellon robotics scientist said, "In our worst nightmare, one of the legs goes down into the earth and just doesn't come back up."[2] The robot was designed to be *statically stable*: with eight legs it would always have more than two on the ground at a time, even while walking. It was engineered to be able to move autonomously over rough terrain, but nothing in its programming allowed it to correct for unexpected slips and falls.

This was not a limitation of Dante II's design, which was state of the art. Robotics had long been plagued by the inability of machines to adapt to complex environments. A robot might easily navigate a simulated office environment, with simple geometric shapes representing office furniture, but it would fail utterly when faced with the complexity of a real office.

But new strategies were being developed. When Dante II took its fall, robotics scientists were taking a new approach to machine design: using biological systems as inspiration. Evolution had already solved many of the problems that robotics was struggling with, so it was natural to turn to the products of evolution for solutions. An insect, for example, has very much the same shape as Dante II, but an insect can climb over exceedingly complex terrain and can even adapt to the loss of one or more limbs, a skill that a robot functioning in a dangerous environment would need to emulate. The research field that developed, at the intersection of biology and robotics, is *biorobotics*, and the falling cat has become a significant subject for research in this field. Cat-turning, in fact, has proven to be a particularly difficult feat for robots to emulate.

Biorobotics can be broadly divided into two subfields of study.[3] The first is *biologically inspired robotics*, in which biological systems are studied in order to build better robots. The second is *biorobotic modeling*, in which robotic models of animals are constructed in order to better understand animal biology.

Both of these strategies existed long before the word *robot* had ever been coined, and even before electricity was harnessed to make powered machines. The original falling cat photographer Étienne-Jules Marey, for instance, made mechanical models of the circulatory system and of flying insects and birds. He used these *schéma*, as he called them, to understand how animals live and move.

Ironically, the first mechanical model of a falling cat—albeit a very crude one—was made by Marcel Deprez, the most vehement early objector to Marey's falling cat photos, in 1894. After coming around to Marey's way of thinking, Deprez published a paper in which he describes his device.[4]

A flat disk is hanging freely from a thread. Carved into the disk are two circular grooves holding metal spheres that can be propelled by springs. When both springs are released, by burning a wire holding them in place, they launch the two spheres around the full circumference of the grooves, and the spheres end at the same place

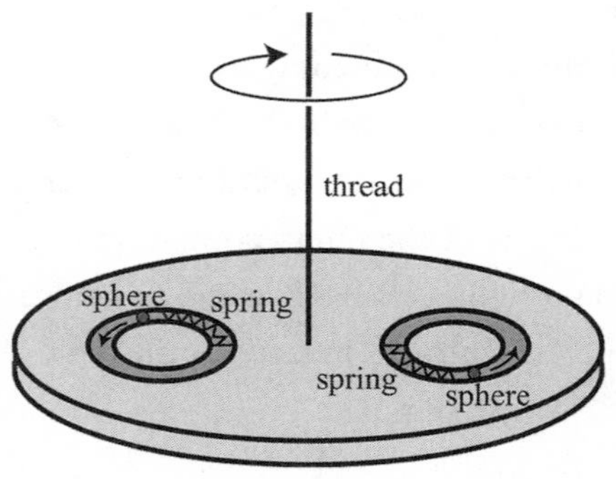

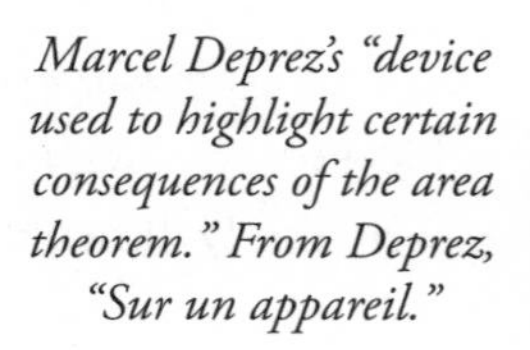
Marcel Deprez's "device used to highlight certain consequences of the area theorem." From Deprez, "Sur un appareil."

they began. Both spheres circulate in the same direction, and by conservation of angular momentum, the entire disk must rotate some amount in the opposite direction. The counterrotation will not be a full 360 degrees, however, because the disk is much heavier than the spheres. The result is that the system ends up in the same internal state that it started with—we will neglect the small change in the spring tension—but it has rotated by a finite amount. In Deprez's test, the device rotated by 40 degrees. This system, he argued, is analogous to a cat, for a cat uses internal motion to end up facing a different direction, even though in the end its body shape is the same as when it began.

Deprez's cat model is a good example of biorobotic modeling; he used a mechanical model to explain the falling cat problem. An example of the other form of biorobotics, bio-inspired robotics, can be found in the "Steam Man" of George Moore, invented around the same time.

Humankind has long been obsessed with the construction of *automatons*, machines that move and behave like living creatures. They may be considered the precursors of modern robots. Moore's Steam Man, discussed in *Scientific American* in 1893, was a powerful and potentially explosive automaton in the shape of a man that could walk at a speed of about four or five miles an hour.[5]

The Steam Man, which was fashioned to look like a marching knight, was powered by a steam boiler in its chest; exhaust came out through the nose. The boiler drove the gears that made it walk. The images are somewhat misleading, however, for they do not show the horizontal bar that the Steam Man was mounted on; this bar was fixed to a rotating platform. The Steam Man marched in circles, and the bar kept him from falling over.

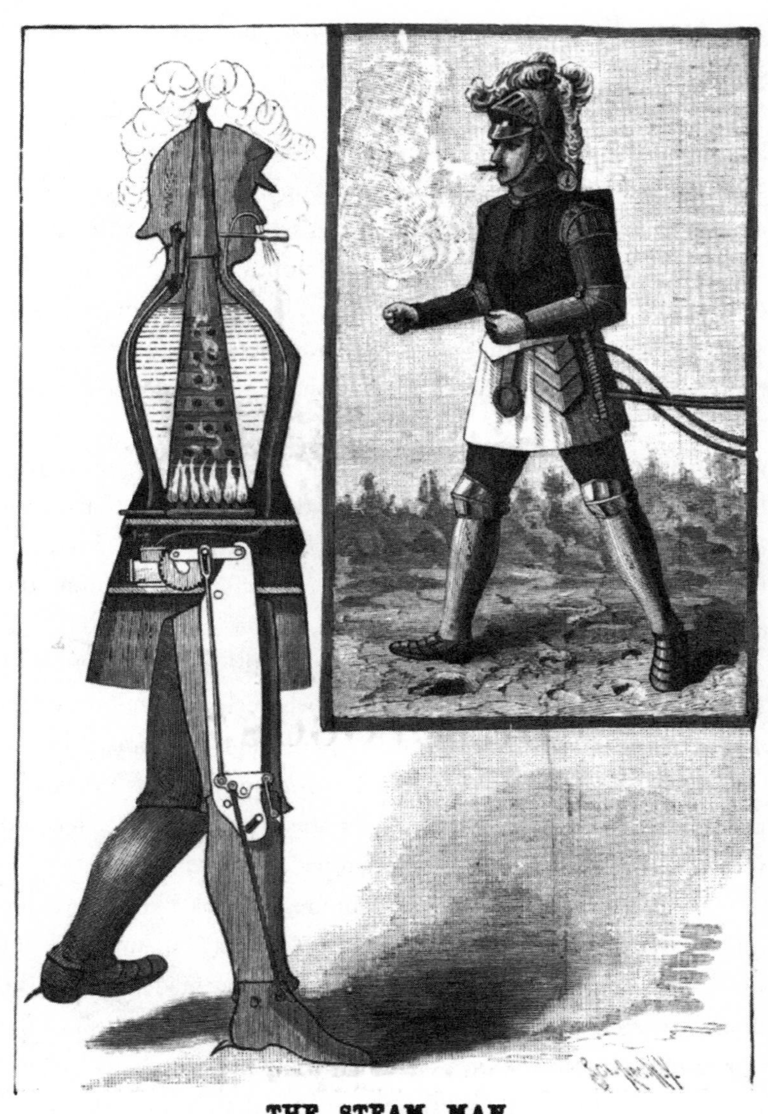

George Moore's Steam Man. From "The Steam Man."

But George Moore did not design his automaton simply for show; he had much bigger ambitions, as *Scientific American* reported:

> For the last eight years the inventor has been at work on a larger steam man, which he hopes to have in operation during the present year. The new one is designed for use on the open streets and is to draw a wagon containing a band. In the upper figure we indicate the method of attachment to the wagon which has been adopted. By the long spring at the side of the figure an elastic connection is secured, so that the figure shall always have its weight supported by the ground.

If Moore's Steam Man shows that humanity has been fascinated by the possibility of artificial life, it also demonstrates that humanity has simultaneously feared that possibility. As planned, Moore made his automaton more mobile following its unveiling, but not everything went as planned, as was reported in the *Washington Standard* in 1901.

> "Hercules, the Iron Man," is a steam mechanical walking man on exhibition at a Cleveland (O.) summer resort. He is eight feet high, and when the oil fire inside him is lighted and steam generated, he walks about, pushing a sort of iron-wheeled cart. He wears a plug hat and a fiendish grin, and puffs exhaust steam through his nostrils. Late one night some of the campers at the park lighted the fire in "Hercules" after the resort was closed and "Hercules'" owner had gone away. The valve had been left open when "Hercules'" fire was put out, and when he got up steam he began to walk about the park. He beat Frankenstein's monster for a while.

> No one knew how to stop him, and he walked all over the park, through the shallow lake, over the tents of the campers and the sideshow tents. Sleepers in his path had to be awakened to get them out of the way, for it was impossible to control the steam man's movements. Inequalities in the ground, trees and other obstructions turned him aside, but could not stop him. He terrorized the park for an hour, but came to grief at the bar. He marched up to it just as though he had money, bumped against it and knocked it over. "Hercules" fell with the bar and alighted on his head on the other side. He stood there on his head, kicking his feet in the air until his steam went down.[6]

This story, which seems highly exaggerated, does capture the problem that has plagued autonomous machines for most of their existence: their inability to adapt to unexpected real-world obstacles. If the article is to be believed, Hercules was thwarted by a bar, just as Dante II was thwarted by a muddy slope.

The fear of artificial life is intimately tied up with the word *robot* itself, which first appeared in the 1921 science fiction play *R.U.R.* (Rossum's Universal Robots), written by the Czech author Karel Čapek. In the play, artificial people—robots—are produced in a factory by the Rossum company. Eventually the robots, who can think for themselves, rise up in rebellion and exterminate almost all of humanity. In the end, the last human, realizing that the robots have developed humanlike compassion, helps them find the lost secret of their reproduction, ensuring that they will inherit the Earth. The word *robot* was derived from the Czech *roboti*, which refers to a serf-like forced laborer.

Fears of a robot uprising aside, the development of some sort of free will, or at least an ability to adapt, is essential for machines to be able to function in any real-world environment. An early pioneer in this line of thinking was the physiologist and roboticist W. Grey Walter, who in 1949 introduced the public to a pair of autonomous robot turtles that he had designed, Elmer and Elsie.

Walter's work began as an experiment in biorobotic modeling. He built his robots to gain insight into the operation of the nervous systems of living creatures. Walter, who was born in 1910 in Kansas City, had studied physiology at Cambridge and then did research on neurophysiology. In 1935 he became interested in electroencephalography—the measurement of electrical activity in the brain—a subject to which he made major contributions over the next few decades, working as director of physiology at the Burden Neurological Institute in Bristol, England.

His work in robotics sprang from his desire to understand how the living brain can implement complex behavior through its components and interconnections. The two tortoise-like robots that he constructed, Elmer (ELectro MEchanical Robot) and Elsie (Electro mechanical robot, Light-Sensitive, with Internal and External stability), were basically shells with wheels, each sporting a periscope-like eye. The number of electrical components in each tortoise was exceedingly small: two radio tubes (acting as neurons), one a light sensor and the other a touch sensor; two motors (one for crawling and one for steering); and two batteries. Here is Walter's description: "The number of components in the device was deliberately restricted to two in order to discover what degree of complexity of behavior and independence could be achieved with the smallest number of elements connected in a system providing the greatest number of interconnections."[7] In short, Walter hypothesized that the complex behavior of living creatures comes as much from the interconnections and interactions of sensory organs and neurons as from the number of neurons present and that quite sophisticated responses could even come from an "animal" with only two neurons.

The results of the tests, at least those that Walter presented, were impressive. The light sensor "eye" of a tortoise would rotate until a sufficiently bright, but not too bright, source was spotted. Then the machine would drive toward that source. When the light got too bright, the machine would switch into a light-avoidance mode and veer off in search of a more hospitable environment. The touch sensor connected to the shell would make the robot change course after a collision was detected, allowing it to successfully maneuver around small walls or even mirrors in its quest to seek light. Even more notably, the tortoise Elsie was designed with a recharging capability. When her batteries ran down sufficiently, her light-avoidance capability diminished, causing her to seek out a brightly lit hutch where automatic recharging would take place.

Walter argued that the complex and unpredictable motions of the tortoise resembled something like free will. In justification, he referenced the philosophical paradox known as *Buridan's ass*, introduced by the French philosopher Jean Buridan in the fourteenth century. In the paradox, a hungry donkey is placed equidistant between two identical stacks of hay. If the donkey is simply a mechanical contrivance, both piles of hay will look equally acceptable, and it will in principle starve to death because it cannot choose the closest, or "optimal," hay pile. An animal with free will, so the argument goes, will have no difficulty in making what is essentially an arbitrary decision.

If Elmer or Elsie is placed between two equidistant light sources, the problem is solved almost trivially because the light sensor rotates in a fixed direction. Therefore, the robot will move toward whichever light it sees first. This, to Walter, was a mechanical demonstration of how living creatures could overcome the paradox: though the lights are equidistant in space, they are observed at different times. The problem of Buridan's ass or, more generally, the problem of robots becoming "stuck" between multiple objectives that are equally desirable, can thus be defeated.

Walter's tortoises, to which he gave the playful species name of *Machina speculatrix*, were not perfect. They achieved the successes they did largely because they were operating in a very simple environment.[8] But they were the earliest robots that were biologically inspired, and they showed how a merging of robotics and biology could result in surprisingly complex machines.

The merging of biology and technology was not restricted to robots. In the late 1950s, researchers began to study biological systems with an eye toward applying the lessons of evolution to new devices and products. The term *biomimetics* was introduced for this strategy of drawing inspiration from nature; in 1960, Jack Steele of the U.S. Air Force coined a now more familiar term for this idea: *bionics*. An early product of this strategy is Velcro, developed by George de Mestral in 1941 after he found hooked seeds caught in the fur of his dog after an outdoor walk. Other examples of biomimetic design include dry adhesive tape, which was inspired by the adhesion of gecko feet to walls, and antireflective surfaces for glass, which were inspired by insect eyes and wings. It has even been suggested that tires with improved and more versatile traction could be developed by mimicking cats' claws.[9] And, as we have noted, the grooming habits of cats are already inspiring new technologies.

For some time, robotics remained focused on task-specific applications. One of the milestones in bringing robotics into practical use was the introduction of the first industrial robot, Unimate. Short for "universal automation," Unimate was the first digitally programmable robot arm. It was built with two goals in mind. The first was to create a machine that could handle dangerous tasks in factories where workers are exposed to toxic substances and hazardous machinery. Unimate's creator, George Devol, saw an opportunity to make the workplace safer by giving the riskiest tasks to robots.[10] The second goal was to reduce the waste produced by obsolete manufacturing machines. By making the robotic arm programmable, its actions could be changed as needed when production methods changed. Devol, together with his business partner, Joseph Engelberger, founded the company Unimation in 1962 to manufacture and sell their machines.

Unimate was perfectly suited for the jobs it was designed for, but it was a stationary robot with a fixed pattern of motion. To make robots even more versatile, stable motion was required, and this demand spurred invention in the field of biorobotics in the 1980s and 1990s. Most early researchers in autonomous robots designed the control center—the "brain" and "nervous system" of the robot—to be separate from the machine itself, as in the volcano-climbing Dante II, whose reactions to the environment were controlled from a station at the edge of the crater, connected to Dante by a tether. Robots that could react well in a crisis, however, would need to have their brain and reflex actions tied tightly to their sensory and motor operations.[11] In short, they would have to be built much more like living creatures.

Inspiration for this approach was not limited to terrestrial animals. It has long been appreciated, for instance, that fish are more efficient and maneuverable than human-made boats. Fish require less energy to propel themselves through the water, they can generate incredible bursts of acceleration while chasing prey, and they can turn suddenly to avoid becoming prey themselves. In the summer of 1989, while chatting with colleagues at the Woods Hole Oceanographic Institution at Cape Cod, brothers Michael and George Triantafyllou realized that there was a great need for efficient robots for deep sea exploration. Fish locomotion, they surmised, could provide insights. By studying a variety of fish, from goldfish to sharks, they found that there is an optimal method of flapping the tail to produce propulsion. Drawing upon their observations, they constructed a forty-nine-inch-long mechanical bluefin tuna. The biological principles that they had discovered worked well for their mechanical model. Significantly, in the closing remarks of their published paper, they raise some profound questions about the swimming of fish.

> Although the dolphin and the tuna are both fast and flex their bodies in similar ways while swimming, there are significant differences in the details of swimming as well. Are they both optimal solutions? If one is better than the other, is the superiority limited to certain situations? More important, as far as we are concerned, is there an even better design than either of them for swimming?[12]

To put it another way, Are the swimming styles of the dolphin and tuna equally efficient and, if so, how can we choose one over the other? The question is, indirectly, a form of the Buridan's ass paradox. It is not an issue for living aquatic creatures, whose methods of choice have been forged by evolution, but it is a significant issue for robotics scientists.

When the robotic tuna took its first swim, researchers at Case Western Reserve University were working on improving terrestrial robot locomotion through biology. They developed several six-legged insect robots (hexapod robots) based on the anatomy and neurology of cockroaches and stick insects, including as many biological inspirations as possible. As the researchers noted, "We tend to err by including more biology than may at first appear to be strictly necessary. The reason for this strategy is straightforward: it almost always pays off. While nature's way may not be the only way, or even necessarily the best way, time and again we have found unexpected benefits to paying close attention to the design of biological systems."[13]

The insect robots were designed with the lessons learned from reflex research of the late nineteenth and early twentieth centuries. Each limb was equipped with what were effectively proprioceptive reflexes: artificial neurons that sent information to their managing pacemaker neuron about the forward or backward orientation of the limb. Furthermore, the different pacemaker neurons of the legs mutually inhibited each other, just like the nerves of antagonistic muscles in living creatures. This inhibition improved the coordination of the legs, preventing adjacent legs from stepping at the same time. A number of other reflexes were built into the system. One was called the elevator reflex: if a forward-swinging leg encountered an obstacle, it would withdraw, raise itself to a higher elevation, and attempt the step again. Another reflex was the "searching reflex": if a leg at the end of its swing did not find a stable foothold, it would search the local area until it did.

In 1994 the researchers tested a stick insect–inspired robot, approximately 50 centimeters (20 inches) long and 25 centimeters (10 inches) tall, on rough terrain. The terrain was a large piece of Styrofoam packing material with elevation changes of roughly 11 centimeters (roughly 4 inches). Styrofoam served as a good test because it was soft and would flex and bend, making an unstable surface similar to the one that Dante II stumbled on. The robot performed admirably. It was able to integrate and apply the elevator and searching reflexes in coordination to move 2 centimeters (1 inch) per second over the rough terrain.[14]

In an interesting callback to the physiological research of the early twentieth century, the same research group also studied the robustness of their robot when neural connections were severed. By introducing "lesions" in various neural connections, they were able to show that their robot maintained a significant degree of functionality when damaged. Even the complete loss of a single leg's function did not seriously impede the robot's motion.[15]

Though the hexapod robots represented an impressive step forward, so to speak, robots need to be even more versatile to survive uncontrolled environments. Real-world terrain has drops that are larger than the robots themselves, so they must be prepared to compensate for unexpected freefalls. This becomes of crucial importance for robots designed to climb smooth vertical surfaces, such as a gecko-inspired robot built in 2007.[16] Here the falling cat problem becomes important again, providing a potential method for robots to right themselves and land on their feet with minimal damage and continuing mobility.

Work on building practical cat robots has lagged behind other projects. Outside robotics, an ingenious mechanical model of a falling cat was introduced by John Ronald Galli of Weber State University in 1995, at roughly the same time that the robotic tuna was swimming and the robotic stick insect was marching. Galli was inspired to study the problem after reading a 1980 article by Cliff Frohlich about the physics of cat twisting and human divers. Frohlich's article describes, without citation, the 1935 bend-and-twist model of Rademaker and ter Braak.[17]

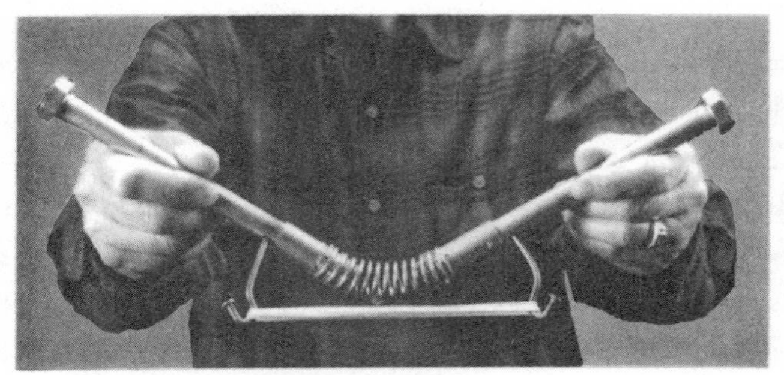

Fig. 2. Energized mechanical spine.

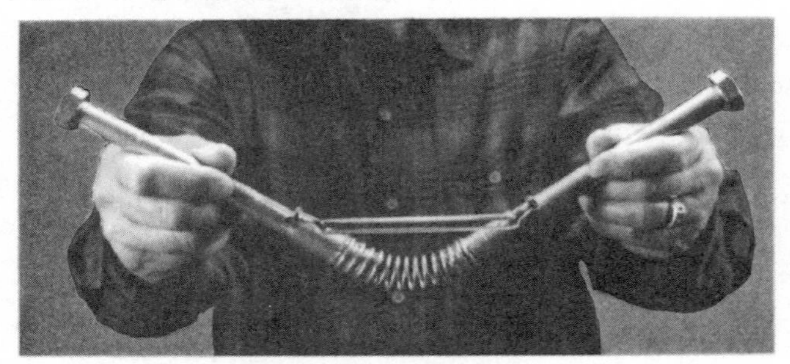

Fig. 3. Energized spine as it is turning over.

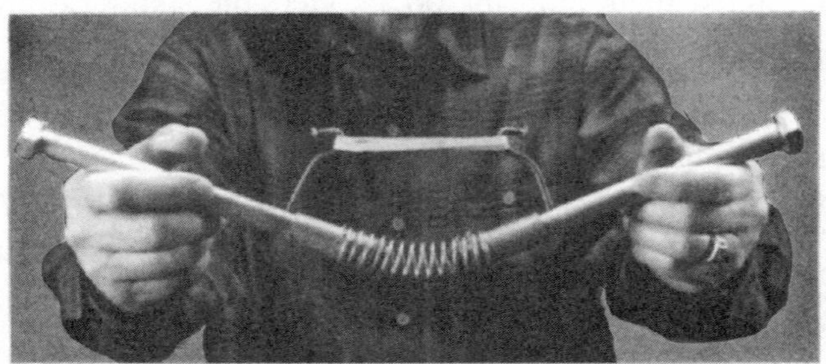

Fig. 4. Mechanical spine after turning over.

John Ronald Galli's mechanical cat held at several stages of its motion. Reproduced from J. R. Galli, "Angular Momentum Conservation and the Cat Twist," Physics Teacher, *33:404–407, 1995, p. 404, with the permission of the American Association of Physics Teachers.*

Galli constructed several increasingly complex models of the cat; the operation of the simplest is shown here. The two cylinders act as the front and rear halves of the cat; the spring, as the flexible spine. A piece of elastic between the two halves of the body acts as an energized muscle; when the cat is released, the tension in the elastic causes the body sections to perform a bend-and-twist, resulting in a 180-degree flip.

Galli's model falls broadly into the category of biorobotic modeling. His device has become a standard tool for physics educators interested in explaining the problem to their students. Until recently an advanced version of the Galli cat was available online for purchase, complete with legs, extra joints in the spine, and a wire kitty face.

To implement a robust cat-turning strategy in a robot is a more difficult proposition. Because a cat can fall from a variety of angles (upside down, on its side, head first, etc.) and can start the fall with or without angular momentum, no simple script will right a cat in every circumstance. A precise motion that causes an upside-down cat to land right-side up, for example, would cause a sideways cat to land on its opposite side. The cat or robot, in falling, has to take into account the exact circumstances of its fall and adjust its strategy accordingly, often within a fraction of a second.

The difficulty in designing a robot to do this comes back to the problem of Buridan's ass. Since a cat can in principle use a number of different methods to right itself, if we introduce the criterion that the cat must flip over in the shortest amount of time, two strategies could end up taking the same amount of time, and the robot could land on its back simply because it couldn't choose. As two researchers note, "The question of a general strategy and a control approach to the problem of reorienting a complex multisegment, multijoint system without the use of net external torques has remained largely unanswered. The difficulty in answering this question lies in the fact that, in general, there is an infinite number of ways a multisegment, multijoint system can be arbitrarily reoriented in space without the use of net external torques."[18]

To make a robot that won't get stuck between two equally good methods of turning over, the engineer must therefore introduce a very precise definition of "good" that allows only one possible method in any specific circumstance. That is why a number of studies of cat-turning have focused entirely on the mathematical solution of the problem. One early paper used photographs of falling cats to study the role of the vestibular system in the control of the falling cat's motion.[19] This work was not focused on robotics, but the results would guide the work of future researchers.

In 1998, Ara Arabyan and Derliang Tsai of the University of Arizona designed an algorithmic control scheme for a falling cat that would allow it to successfully turn over. This scheme, like the one for the earlier hexapod robots, was decentralized; the actuators controlling the joints would interact with each other and provide feedback, like proprioceptive reflexes. The authors imposed several constraints on the cat's motion, as originally suggested by Kane and Scher, to restrict the difficulty of the problem that needed to be solved and implicitly avoid Buridan's ass. One of their simulated falling cat solutions captures the glory of late 1990s computer animation. As the authors point out, their computer-generated solution to the falling cat problem matches actual photos of falling cats quite well.[20]

Work on the mathematics of the falling cat problem continued into the new millennium. In 2007, Chinese researchers used a technique called nonholonomic motion planning to find solutions for the falling cat problem. In 2008, Israeli researchers introduced a curious "square cat" model—in which the cat is four equal-length

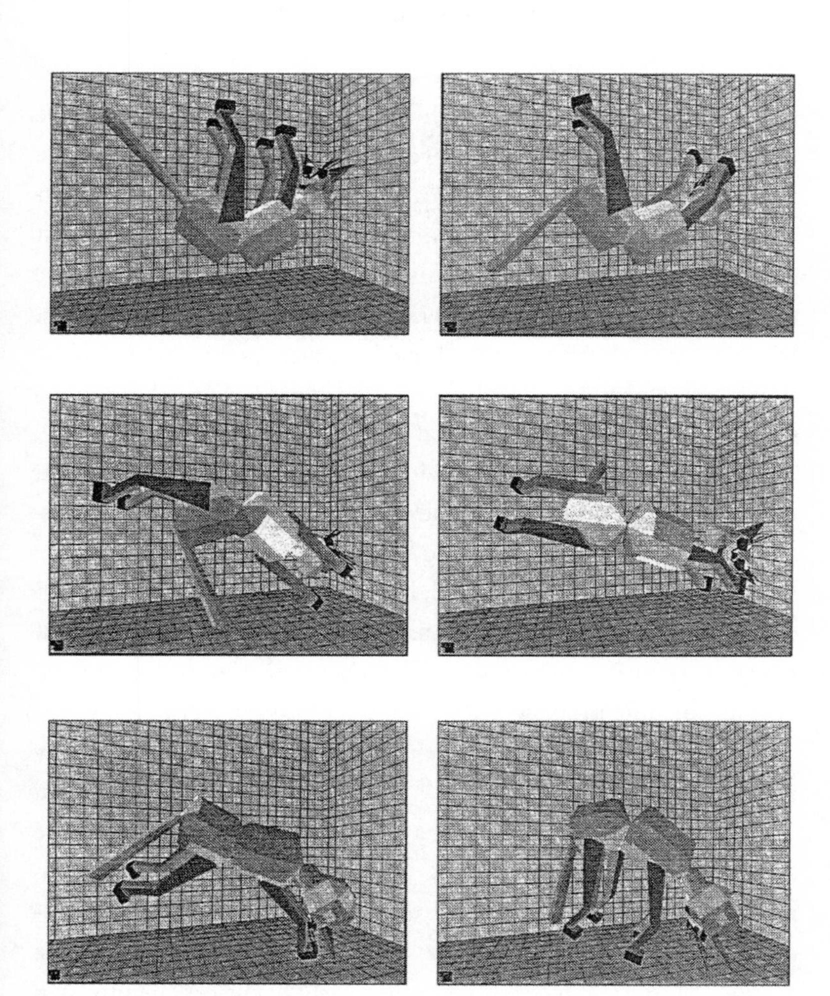

Falling cat simulated by Ara Arabyan and Derliang Tsai. (For similar images see the "Money for Nothing" video by Dire Straits.) From Arabyan and Tsai, "A Distributed Control Model for the Air-Righting Reflex," reprinted with permission from Springer Nature, copyright © 1998, Springer-Verlag Berlin Heidelberg.

rods connected by flexible joints—to elucidate some of the deeper mathematics of the problem. In 2013, Richard Kaufman of the University of Massachusetts Lowell introduced an "electric cat," a simple mechanical model of a cat performing the bend-and-twist maneuver. One of Kaufman's conclusions is that the bend-and-twist is more than sufficient to account for the cat's ability and that the tuck-and-turn method of Marey is at best a secondary contribution. In 2015 a different group of Chinese researchers applied a sophisticated mathematical formalism, the Udwadia-Kalaba equation, to study the dynamics of a falling cat.[21]

In most of this later work, with the exception of the electric cat model, the emphasis has shifted away from explaining how a cat performs its movement to explaining how to use mathematics to get the same result. The cat, it is implied, has already figured out how to flip over in an optimal way; the goal of the mathematician is to figure out how mathematical rules might approximate this evolution-derived decision-making process.

Cat-turning appears to be the most sophisticated method by which any animal self-rights itself, but other creatures have adopted simpler techniques, and these have been explored in much more detail by robotics researchers. A 2011 review of self-righting techniques presents four options, many of which we have seen previously.[22] "Modification of angular momentum prior to leaving the ground" is comparable to the original, insufficient explanation that Maxwell and others had for the falling cat: a robot or animal can adjust its own angular momentum to start rotation before it falls. "Body reorientation through limb motion" refers to techniques such as the Air Force's lasso maneuver: by circling the arms, the body can be counterrotated. "Twisting with no initial angular momentum" refers to catlike flipping, both bend-and-twist and tuck-and-turn.

The fourth technique, perhaps the least desirable, is the most practical: "a posteriori self-righting," that is, righting *after* hitting the ground. Some animals can attempt to correct their orientation once one limb hits the ground, but before the whole body does, by using that limb for leverage.[23] Other creatures in nature have concocted spectacular methods for getting up after landing upside down, including beetles.

> In the event that they land on their backs, some species of beetles (such as Eucnemidae) are known to arc their body, storing elastic energy, before explosively releasing it in a jump aimed at self-righting. Other species, such as the Histeridae [clown beetles], open their elytra (the hardened protection of their hindwings) into flight position against the ground, then explosively shut them, resulting once again in a self-righting jump.[24]

In a sense, this latter technique is a do-over: if a beetle lands on its back, it throws itself back into the air and tries again to land upright.

Some insects, owing to their small mass, apparently don't need a strategy for self-righting. Research on larval stick insects indicates that aerodynamic forces alone are enough to cause them to flip over; in essence, the "wind" they experience while falling causes them to flip.[25] This phenomenon, remarkably, is very much what was suggested by Antoine Parent in 1700 as the way that cats turn over. Though it is wrong for cats, it is apparently accurate for some species of insect.

For other animals, anatomy allows for much easier turning strategies. Because a lizard has a tail that is comparable in size to its body, it can use the "propeller tail" strategy, first suggested by Giuseppe

Peano, to flip its body over. In 2008 researchers at the University of California, Berkeley, analyzed the flipping of the flat-tailed house gecko, *Hemidactylus platyurus*, in order to design a bio-inspired self-righting robotic lizard. They built their prototype with the same size and shape as the gecko-inspired sticky climbing robot of 2007. The results were impressive: the robot prototype flipped 180 degrees in three-tenths of a second, in agreement with their models and quite sufficient for most self-righting needs.[26]

These creatures, and the robots they inspire, can use their tails for control in jumping as well. In a paper that may have the greatest title of all time, "Tail-Assisted Pitch Control in Lizards, Robots and Dinosaurs," the Berkeley research group studied the leaps of

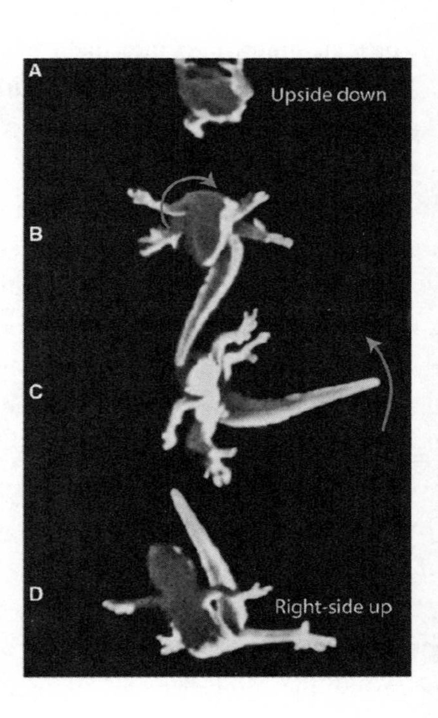

A self-righting gecko. From Jusufi et al., "Active Tails Enhance Arboreal Acrobatics in Geckos," PNAS, *105:4215–4219, 2008, copyright © (2008) National Academy of Sciences, U.S.A.*

Agama lizards and used that data to improve the stability of robot jumps.[27] A robot that launches off a ramp Dukes of Hazzard–style will tend to land face first because gravity starts to pull the front of the robot down before the back of the robot completely leaves the ramp. By swinging its tail upward, however, the robot (or lizard) raises its front end, again by conservation of angular momentum, so that it safely lands flat. The researchers also used their results, and paleontological data, to speculate on pitch control in the cinematically infamous dinosaur *Velociraptor mongoliensis*. They noted, "Despite previously proposed limitations of passive tails, small theropods like Velociraptor with active tails might have been capable of aerial acrobatics beyond even those displayed by present-day arboreal lizards." If Velociraptors no longer seem scary enough, imagine them chasing their prey like parkour-skilled acrobats.

Though a cat's tail is much less effective than a lizard's in such maneuvers, experimental research has demonstrated that a cat also uses its tail for balance.[28] Again, high-speed photography came into play. Cats were filmed walking across a narrow beam that was shifted suddenly to the side while they were on it. Video footage shows that cats swing their tails to counterbalance the unexpected motion.

Other creatures have evolved even more unusual pitch adjustment techniques, and these, too, have been adapted for robotics. Jumping spiders in the salticid family have been found to attach a dragline of silk to their perch before launching into the air; by controlling the tension in the line as they extend it, they can adjust their pitch in order to land flat. In 2015 researchers at the University of Cape Town demonstrated that a robot could use the strategy.[29] Their robot, LEAP (Line-Equipped Autonomous Platform), was built from a chassis of LEGO Technic bricks to keep it as light as possible; the final machine weighed 88 grams (3.1 ounces). This robot determined by itself when to activate its line control by using its own version of a vestibular system: an accelerometer. While on its launching platform, the robot experienced the normal force of gravity; when it was launched, the disappearance of gravity triggered it to begin dragline control, much the way a cat's righting reflex is activated by the weightless sensation.

The more researchers study animal motion, the more they find distinct methods of aerial self-rotation. In a paper that may have the second-greatest title of all time, "Aerial Maneuvers of Leaping Lemurs," Donald Dunbar of the University of Puerto Rico investigated the unusual mid-air reorientation of ring-tailed lemurs.[30] A lemur, which often takes off while high in a tree facing the trunk, is able to launch to another tree and land facing its trunk. In this case, the lemur uses two strategies: it begins its rotation while still holding onto the tree ("modification of angular momentum prior to leaving the ground"), and it uses its tail to adjust its rotation prior to landing ("body reorientation via limb motion").

Even winged creatures take advantage of unusual self-orientation techniques. In 2015 researchers at Brown University demonstrated that Seba's short-tailed bats and lesser dog-faced fruit bats use control of their wing inertia to perform remarkably fast aerial maneuvers.[31] The strategy may be considered analogous to the tuck-and-turn model of cats, where the two wings take the place of the two body sections of the cat. By tucking in one wing, the motion of the other wing has an oversized effect on the bat's rotation. The researchers suggest that this knowledge will help improve the performance of aerial robots.

But have any actual, physical models of robotic cats been constructed? As far back as 1992, Japanese researchers studied the bend-and-twist model of cat-turning using robotics; theirs may well be the first actual robot based on the falling cat principle. The original paper was in Japanese; in 2014, one of the authors, Takashi Kawamura of Shinshu University, published a shortened English description of the work.[32] Their model of the cat is somewhat similar to Galli's mechanical cat, consisting of two cylinders connected at a flexible joint, but it uses an active control scheme. The "muscles" of the cat are actuators powered by air pressure, which allows active control of the robot while it is in freefall. The Japanese researchers' motivation was not, however, to design a versatile self-righting robot but to test the bend-and-twist hypothesis.

Most of the work on robotic cats is recent and in its initial stages, largely, it seems, because of the difficulty of designing a reliable control system. In 2013, researchers at the University of Adelaide made a simulation of a falling cat robot.[33] Faced with a variety of proposed strategies for cat-turning, the Australian group focused on designing a robot that could implement Marey's original tuck-and-turn model. Their simulation, shown here, predicts a cat turning over in a little over a half-second. The researchers have plans to build a working prototype.

In 2014, Karen Liu's group at the Georgia Institute of Technology managed to build a cat-inspired robot that is able to dynamically adjust its position in the air to right itself. The robot does not look much like a cat. It was built in three hinged sections that can bend independently of one another to control orientation, again, like Kaufman, using nonholonomic motion planning. Their robot was not ready for the speed and impact of true freefall, so the researchers tested it by sliding it down a tilted air-hockey table, with promising results. The work gained national attention, even though some articles presented the results tongue-in-cheek: "So in the distant future, when you see a terrifying robot leap from a cliff above, you can blame the cat."[34]

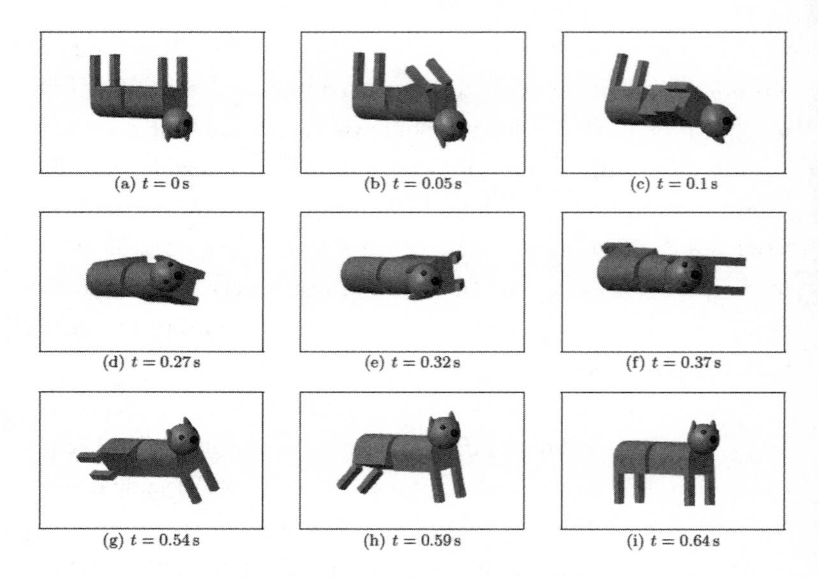

Simulations of a robot employing the tuck-and-turn model. From Shields et al., "Falling Cat Robot Lands on Its Feet"; figure reprinted with permission from the authors.

Other groups made significant progress on the robotic falling cat problem in 2017. In a collaborative project, researchers in Britain and Iran designed simulations of cat robots of increasing complexity—with two body sections, three body sections, and eight body sections—and developed a control system for the robot cat that would avoid "singularities"—that is, the problem of Buridan's ass. A prototype is planned.[35]

A falling robot developed by Morgan Pope and Günter Niemeyer at Disney Research in 2017 is worth mentioning.[36] Their machine doesn't look at all like a cat. It looks like a circuit-festooned brick, is appropriately named Binary Robotic Inertially Controlled bricK, and follows Maxwell's vision of changing orientation. The brick is

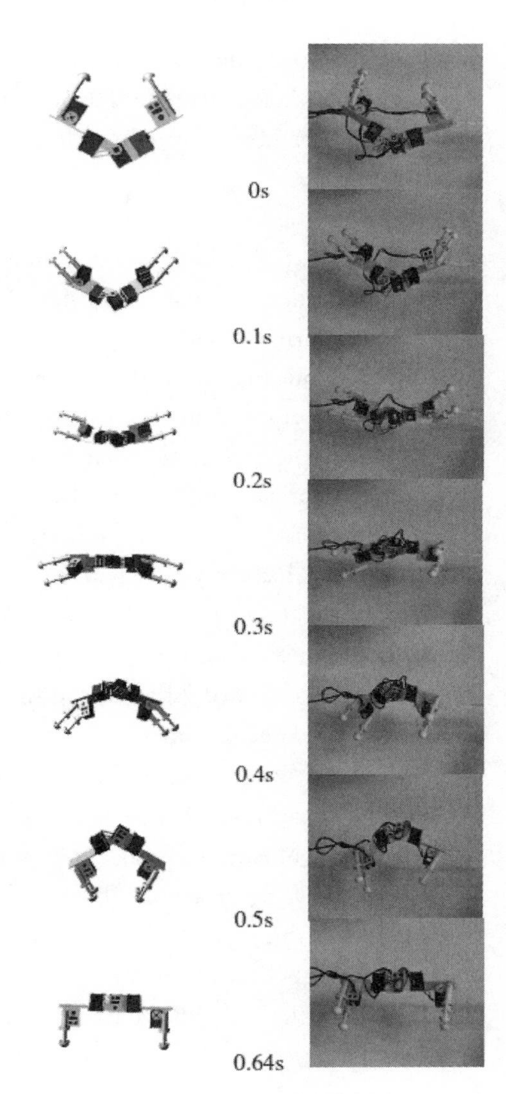

A robot performing a bend-and-twist with "swing legs," shown in simulations and experiments. Reprinted from Zhao, Li, and Feng, "Effect of Swing Legs on Turning Motion," copyright © 2017, IEEE, with permission from IEEE.

given a large horizontal rotation before falling, and it controls its moment of inertia internally to change its rate of rotation. The robot was shown to be able to orient itself to successfully fall through a brick-shaped slot.

What may be the first actual electronics-based robot prototype of the falling cat that looks vaguely like a cat was built by three Chinese researchers, Jiaxuan Zhao, Lu Li, and Baolin Feng.[37] The model of cat-turning they use is primarily based on the bend-and-twist method, but they allow the legs of the robot to swing freely to optimize the motion. This robot itself, though impressive, appears to be following a prerecorded script and not evaluating the optimal motions for landing on the fly, the way a real cat does. In falling cat robotics a complete unification of the mathematics and the mechanisms has not yet been achieved.

Ordinary walking and running robots have, however, advanced dramatically. Since 2013, Boston Dynamics has been developing a humanoid robot named Atlas that possesses amazing amounts of coordination. A large specimen, Atlas is nearly six feet tall and weighs 330 pounds. In 2017, Boston Dynamics released a video of Atlas jumping on boxes and even performing a backflip. A year later it released a video of Atlas running across grass and on uneven terrain. As one online writer reported, "Boston Dynamics' Atlas robot can now chase you through the woods."[38] Apocalyptic announcements aside, Atlas shows how far robots have evolved from the time Dante II struggled to climb out of a volcanic crater on six legs. Still, the fear of robots hasn't changed much since Atlas's steam-powered cousin Hercules appeared one hundred years earlier.

Not every robot being built is a potential threat to humanity. Engineers at the toy company Hasbro are collaborating with researchers at Brown University to improve their initial designs for a robotic cat companion called ARIES (Affordable Robotic Intelligence for Elderly Support).[39] This robot, which looks like a cat, purrs like a cat, and meows like a cat, is designed as an inexpensive companion and assistant to seniors. ARIES has limited motions that simulate a living cat's, like rolling over for belly rubs, and can be programmed to remind owners of impending doctor's appointments and times to take medicine. Being cute and fluffy like a cat is yet another bio-inspired characteristic that modern robots are emulating.

11

The Challenges of Cat-Turning

> Once upon a time, so the story goes, some blind men were sitting by the wayside. They heard that the elephants were coming. They expressed a desire to see them. One blind man felt an elephant's leg and said that an elephant looked like a column. Another felt his trunk and declared that it looked like a rope. A third man felt his ear and was certain that it looked like a fan. The fourth man felt its tail and was confident that an elephant looked like a snake.[1]

In reviewing the recent research on falling robotic cats, it is striking that there still remains significant disagreement as to exactly how a cat turns over in freefall. Some of the researchers looked to the tuck-and-turn method to design their robots; others, to the bend-and-twist model. The different opinions of roboticists mirror those of physicists. Robotics researcher Takashi Kawamura described the confusion: "Interestingly, explanations in physics and dynamics textbooks that deal with feline self-righting remain contradictory and ambiguous."[2]

It is no doubt surprising that a question so seemingly mundane as how a cat turns over in freefall has intrigued and confused even scientists equipped with cutting-edge theories and techniques for over a century, from the time of Marey's falling cat photographs right up to the present day. In a world where we have mastered the atom, built a global Internet, and sent people to the Moon, how can scientists still struggle to understand and replicate the motion of a cat?

The answer is, in part, that the strategy traditionally used by physicists to analyze problems is not completely in line with the way nature, in the form of evolutionary processes in living creatures, actually solves them. A good example of the way physicists are trained to think can be seen in the work of Isaac Newton. Newton took a bewildering array of observations about the motion of planets, comets, and terrestrial objects and unified them into a single theory of gravitation that could explain all of them (with the aid of his laws of motion and a lot of mathematics). The idea of taking complicated observations about nature and reducing them to their simplest form has been a guiding principle of physics ever since. We have already noted how, in the 1860s, pioneering cat-dropper James Clerk Maxwell showed that the seemingly distinct phenomena of electricity, magnetism, and light could all be explained as a single fundamental force, electromagnetism. A century later, in the 1970s, researchers further showed that the weak nuclear force that governs the decay of unstable elementary particles can be connected with electromagnetism, and together they can be explained as a single fundamental phenomenon, the *electroweak interaction*. Now particle physicists are seeking, both theoretically and experimentally, a *grand unified theory* that can connect the electroweak interaction with gravity and the strong nuclear force and show that they are all different aspects of another single fundamental force.

Physicists, then, have a long history of taking complicated physical observations and trying to boil them down into a single phenomenon. This is not always the case—as the problems that physicists study have become more difficult, their strategies have evolved—but physicists have an institutional instinct to hunt for a single "cause."

Nature is interested not in simplicity but in efficiency. In nature, there is no benefit in having the simplest solution to a problem, only in having the best one, which may involve several behaviors or motions combined together. We can see this in the number of distinct cat-turning strategies uncovered so far, of which—discounting Antoine Parent's inaccurate hypothesis—there are four:

1. "Falling figure skater," by James Clerk Maxwell (c. 1850): A cat that is already rotating when it falls can alter its rotation speed by pulling in or extending its paws, which changes its overall moment of inertia.
2. "Tuck-and-turn," by Étienne-Jules Marey (1894): By selectively pulling in one set of paws or the other, a cat can alter the moment of inertia of that section of its body, allowing it to rotate first one half, then the other, without significant counterrotation.
3. "Bend-and-twist," by Rademaker and ter Braak (1935): By bending at the waist, a cat can counterrotate the two sections of its body, canceling out their angular momenta.
4. "Propeller tail," by Giuseppe Peano (1895): By rotating its tail like a propeller in one direction, a cat can make its body rotate in the other direction.

Which of these strategies is the One True Strategy for cat-turning? Many physicists, like the proverbial blind men examining the elephant, have singled out a particular aspect of the cat's complicated motion, ignoring all others, and declared that aspect

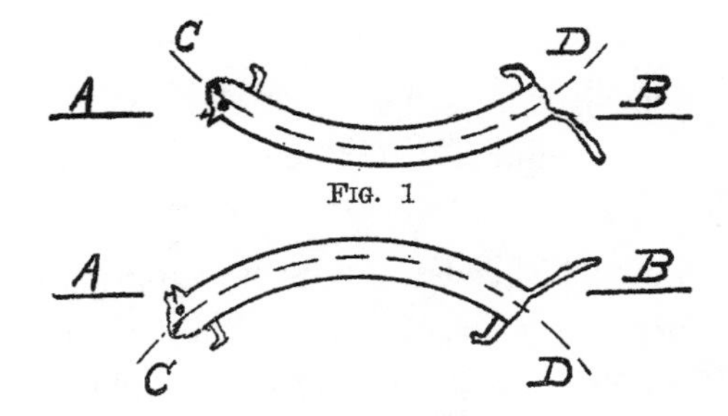

The model of a falling cat provided by W. S. Franklin. From Franklin, "How a Falling Cat Turns Over in the Air."

the "correct" one. Sequences of photos often act as a Rorschach test of sorts for physicists, with each observer seeing something different from the last.

Selective vision on cat-turning goes back almost to the time of Marey. In 1911, W. S. Franklin published a letter in the journal *Science* in which he presented an explanation for the falling cat's motion as given to him by J. F. Hayford. In the letter, Franklin provides an illustration of the cat's motion and an introductory description: "There are two simple types of motion of the cat's body which give spin momentum around the axis AB, namely, (a) a rotation around AB as an axis of the cat's body as a rigid structure, and (b) a sort of squirming motion in which each part of the cat's body rotates about the curved line CD." This amounts to a crude description of the bend-and-twist model that would be championed some twenty years later by Rademaker and ter Braak. But Hayford's explanation was shot down in short order in another letter to the editor, this one by J. R. Benton. Citing a book that contains Marey's photographs and Marey's explanation, Benton states: "The explanation offered by Professor Hayford, although a possible one, accordingly does not agree with the actual performance of a cat, as observed by photography."[3] Perhaps because of this criticism, Hayford's explanation did not gain any traction, and it would take another two decades before the bend-and-twist would become a serious contender for the One True Strategy.

Arguments based on photographic analysis are still made today. In a paper I submitted to a physics journal concerning the falling cat, I used the bend-and-twist as a simple model for the cat's motion. One negative referee report came back with this criticism: "When I study the YouTube movies of falling cats I do not see this kind of motion." The very first photographs of a falling cat, in 1894, did not solve the problem but deepened its mystery. So it goes today.

Among the many researchers into the falling cat problem, only one seems to have taken a much more nuanced view of its complexity: the London physiologist Donald McDonald, working at St. Bartholomew's Hospital Medical College. He published his first paper on the subject in 1955, working at about the same time that the U.S. Air Force was similarly concerned with the problem. McDonald explained his interest:

> The righting reflexes have a time-honoured place in the physiology lecture syllabus and a cat is always, equally traditionally, dropped upside-down to illustrate them. For though the way a cat does it has long been a physiological riddle, Magnus described it in terms of the head and body reflexes he originally discovered—and this is now repeated in all the textbooks. I suppose I must be slow-witted, for I confess that I could never see what I was supposed to see.[4]

Out of curiosity, McDonald decided to explore the problem himself. At first, he attempted to film the falling cat using a motion picture camera filming at sixty-four frames per second, but this was not fast enough to clearly observe the actions of the cat. So he contacted a colleague, John Holland, who specialized in high-speed cinematography, and together they filmed the falling cat at an astounding fifteen hundred frames per second. At that speed the film was running through the camera at sixty miles per hour. A staggering amount of film must have been used to capture an event lasting a fraction of a second.

What did they see? As McDonald wryly comments, referring to the explanations of Magnus, Marey, and Rademaker and ter Braak, "One may well wonder how three different observers could see such varying pictures. To know the answer would be to understand a great deal about the progress of research and the psychology of researchers."

In the end, McDonald saw no evidence for Magnus's screwlike rotation of a cat—which is not surprising, since Magnus's explanation violated angular momentum conservation. McDonald did, though, point out that Marey, Rademaker and ter Braak were at least partially correct. The cat does bend and twist, as the latter two argued, but it also turns at the waist and extends and retracts its paws, as Marey noted. McDonald also observed that a cat will rotate its tail, "often in opposite direction to the turn," as Peano had suggested. However, McDonald does not seem to have understood the physics of angular momentum well enough to judge the usefulness of the tail; he suggests that a cat may use a "fluffy tail" to push against air resistance, or use it for pitch control, as would be done in robotics decades later.

But McDonald recognized what others studying the problem did not: that a cat is not forced to choose a single method of righting itself but can use all the options available to optimize the effect. Any scientist approaching the problem of cat-turning who assumes a single underlying strategy is therefore bound to be confused. The argument between tuck-and-turn and bend-and-twist has gone on for so long because a researcher can find evidence of either mechanism in the cat's motion.

This sort of difficulty does not arise only when physicists investigate living creatures. A number of seemingly mundane physical effects have eluded easy explanation for years, even decades, because there are a multitude of possible explanations, as well as because designing experiments to test those explanations is difficult. As with the cat problem, the effects could also have more than one contributing factor.

The Chandler wobble of the Earth, discussed earlier, is one example. Though physicists quickly realized that the non-rigidity of the Earth causes the wobble, research still continues, over a century after its discovery, into several significant factors contributing to the wobble.

One other example is worth exploring briefly. In 1969, a Tanzanian student, Erasto Mpemba, and a physics professor, D. G. Osborne of University College, Dar es Salaam, published a remarkable paper in the journal *Physics Education*. Simply titled "Cool?," it presents evidence from both Mpemba and Osborne that under some circumstances, boiling-hot water can freeze faster than an equal quantity of room-temperature water.[5] The publication of "Cool?" sparked a scientific mystery and controversy that is ongoing fifty years later.

Mpemba had no such aspirations when he made his discovery. In secondary school in 1963, he was simply interested in making ice cream with his classmates, which involved boiling the ingredients, letting them cool to room temperature, and putting the concoction into the refrigerator. Space was limited in the refrigerator, however, and one time Mpemba put in his boiling mixture at the same time his classmate put in his cooled version. Mpemba was astonished to find that his ice cream had frozen first; when he asked his teacher about this counterintuitive result, though, he was met with ridicule. Fortunately, Osborne visited Mpemba's school and, after being questioned by Mpemba, agreed to do experiments himself.

Osborne was surprised by the results. "At the University College in Dar es Salaam I asked a young technician to test the facts. The technician reported that the water that started hot did indeed freeze first and added in a moment of unscientific enthusiasm: 'But we'll keep on repeating the experiment until we get the right result.' "[6]

Mpemba was not the first person to suggest that hot water can sometimes freeze faster than cold. Observations date back over two thousand years. In Greece around 350 BCE, Aristotle wrote:

> The fact that the water has previously been warmed contributes to its freezing quickly; for so it cools sooner. (Hence many people, when they want to cool water quickly, begin by putting it in the sun. So the inhabitants of Pontus when they encamp on the ice to fish (they cut a hole in the ice and then fish) pour warm water round their rods that it may freeze the quicker; for they use the ice like lead to fix the rods.) Now it is in hot countries and seasons that the water which forms soon grows warm.[7]

Centuries later the natural philosopher Francis Bacon, in his 1620 book *Novum Organum* (New Instrument of Science), stated that "water a little warmed is more easily frozen than that which is quite cold." In 1637, the alleged cat dropper René Descartes published *Les Météores* (Meteorology), an appendix to his famous *Discours de la méthode*, in which he notes, "We can also see by experiment that water which has been kept hot for a long time freezes faster than any other sort."[8]

Since Mpemba's ice-cream-making observation, there have been many follow-up experiments, some showing the effect, others not showing anything at all. A great difficulty in answering the question Does the Mpemba effect exist? is that many hypotheses have been proposed, and multiple effects may contribute, as in the case of the falling cat.

Here are some of the hypotheses that have been put forward to explain the Mpemba effect.[9]

- *Convective heat transfer*. When a liquid is heated, it can form convection currents that rapidly bring the hot liquid to the surface, where the heat is lost by evaporation. Osborne noted that convection will keep the top of the liquid hotter than the bottom even when the temperature drops to match that of an initially cold liquid that is not cooled by convection. The drop in temperature results in a faster rate of cooling, which could explain Mpemba's observation.
- *Evaporation*. A boiling or very hot liquid will lose some of its mass to evaporation. With a lower mass, it will cool faster, possibly giving a boost to the Mpemba effect. Osborne already noted, however, that evaporation alone could not account for the entire rate of cooling of the hot liquid.
- *Degassing*. In 1988 a Polish research group successfully observed the Mpemba effect and noted that the effect depended strongly on the amount of gas dissolved in the

water. When the water was purged of air and carbon dioxide, the time for it to freeze became proportional to the starting temperature. The researchers suggest that the presence of gas was slowing the rate of cooling substantially. The heated water, having been purged of gas, could cool more quickly.[10]

- *Supercooling*. In 1995, German scientist David Auerbach proposed that the Mpemba effect could be explained by "supercooling" and performed experiments to back up the assertion. When a liquid remains a liquid below its normal freezing point, which only occurs for a very pure liquid kept very still, it is supercooled. Auerbach suggested that cold water will supercool to a lower temperature than hot water, thus giving the hot water an edge. In a series of experiments around 2010, James Brownridge of the State University of New York at Binghamton tested the supercooling hypothesis and observed the Mpemba effect successfully in twenty-eight out of twenty-eight attempts.[11]
- *Distribution of solutes*. In 2009, J. I. Katz of Washington University suggested that solutes present in the cold water can slow the freezing process, as was suggested earlier, in connection with gas, but also that those solutes get driven from the freezing water into the as-yet-unfrozen water, slowing the cooling process further.[12]

Other papers, and other explanations, exist. The variety of possibilities makes it difficult to isolate the Mpemba effect—or even, in many cases, to produce it reliably. If the Mpemba effect, like cat-turning, depends on more than one distinct mechanism, designing a controlled experiment that tests only one mechanism would likely see no effect. Another challenge is the difficulty of defining, rigorously, the term *freezing*. Does the liquid have to be frozen solid to count as frozen in a Mpemba experiment, or is the first appearance of ice sufficient?

All of these questions might have become moot after researchers at Cambridge University and Imperial College London regrettably concluded in 2016, after an experimental study, that they could see no evidence for the Mpemba effect at all. But, in a twist worthy of a strange phenomenon with a dramatic history, in 2017 two research groups independently demonstrated, *theoretically*, that it is possible for thermal systems to undergo the Mpemba effect. Their work may well keep the controversy alive for another generation of scientists to explore.[13]

Erasto Mpemba himself did not follow up on the work. Instead he went to get a diploma at the College of African Wildlife Management in Moshi. After further education in Australia and the United States, he became the principal game officer for the Tanzanian Ministry of Natural Resources and Tourism. In that role he worked on wildlife management and conservation, no doubt interacting with cats of a much larger size than those discussed in this book. In 2011, now retired, he gave a TEDx talk in Dar-es-salaam on his amazing discovery and his life.

While working as a game officer, Mpemba is not likely to have seen lion- or tiger-turning. There does not seem to be any published research on the subject, but a very unscientific search of online videos suggests that lions and tigers do not have any such reflex. When they get in trouble, they hang vertically from a tree, then drop down on their hind legs. Some smaller wild cats have that capability, however. In a high-speed video by the BBC, an African caracal clearly performs both bend-and-twist and tuck-and-turn maneuvers as it drops toward the ground. In another video, when a leopard tumbles from a tree with its dinner in tow, it is clearly swinging its tail, propeller-like, on the way down.[14]

So there is still room for further scientific research on cat-turning. It is worth noting that Donald McDonald continued to explore the trickiness of cat maneuvers into the 1960s and then expanded that work in a direction that in hindsight should have been obvious: to high diving. High diving began as a sport in Scotland in 1889—only a few years before Marey's famous cat photographs—and quickly evolved into a popular event, with "high-fancy diving" appearing in the Olympics in 1912. Divers can perform complicated twists and turns in the air before hitting the water, all clearly initiated by local twists and rotations of the body.

McDonald was able to test his idea, proposed in 1960, that humans could perform such catlike maneuvers. As he noted in a follow-up article, "As a result of the work on cats I was approached by Mr. Wally Orner who has been responsible for training Brian Phelps, the brilliant diver who won a Bronze medal at the last Olympics. We filmed some simple experiments which showed quite clearly that Mr. Phelps is perfectly capable of doing at least a 360° twist in the air without any help from the [diving] board."[15] To test whether Phelps had any angular momentum to begin with, McDonald instructed him to make a simple jump off the board and twist only when given a shouted command. Very much like a cat, Phelps could make a 360-degree turn in roughly half a second. In additional experiments, Phelps mimicked the cat more directly and hung upside down underneath the board and attempted to turn around after letting go; an illustration of the result is shown here. It appears there that Phelps is performing the Rademaker and ter Braak bend-and-twist and is even

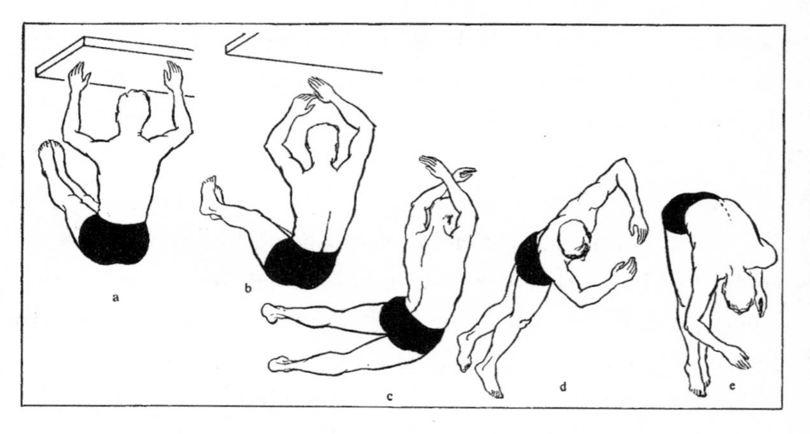

A man doing a "cat-turn." After (a) the initial release, he (b) performs a bend-and-twist until he is (c) sideways and bent left. Then he (d) bends right and (e) continues the bend-and-twist to complete the turn. From McDonald, "How Does a Man Twist in the Air?," fig. 1. Copyright © 1961 New Scientist Ltd. All rights reserved. Distributed by Tribune Content Agency, LLC.

applying the more complicated side-tilt maneuver described by Kane and Scher.

McDonald is not the only author who connected cats and acrobatic sports. In 1974, H. J. Biesterfeldt hypothesized that aerial twists performed by gymnasts are often achieved using what we would now recognize as the bend-and-twist technique. In 1979, when Cliff Frohlich attempted to clear up the mechanism by which springboard divers perform their twists, the falling cat played a role as an illustrative example, as did the twisting astronauts in Kane and Scher's work. In 1993, M. R. Yeadon also cited the cat in a discussion of doing aerial twists while somersaulting. In 1997, Jesús Dapena looked at the contribution of "catting" to the twisting of athletes in the high jump.[16]

These examples notwithstanding, it would be an exaggeration to say that cat-turning has played a major role in the study of sports physics, but it has served to show the remarkable possibilities of human rotation. And clear examples are still evidently needed. As Frohlich noted in his 1979 paper,

> Recently, all the graduate students, postdoctorates, and faculty in the Physics Department at Cornell University were given a questionnaire with specific multiple choice questions about the physical possibility of performing certain somersaulting and twisting stunts. . . . Nevertheless, of the 59 physicists who responded to the questionnaire, 34% incorrectly answered the first question, and 56% incorrectly answered the second—an astonishingly high proportion of misses for multiple choice questions.[17]

This confusion is reminiscent of the response to Marey's photographs of the falling cat almost one hundred years earlier. Even in modern times physicists can be led astray when looking at complicated situations, even those involving simple laws of physics.

So where does this leave us in explaining how a cat turns over? Examination of the evidence, with all models in mind, strongly indicates that the bend-and-twist must be the dominant mechanism used, with refinements provided in the 1969 model of Kane and Scher. But the evidence also suggests that a cat probably uses some mixture of the four models described earlier. None of these options are mutually exclusive, and they can combine readily. A cat can use the bend-and-twist method, but extend its back paws and tuck its front paws so that the front of its body turns right-side up more quickly. It can counterrotate its tail to speed up the rotation of the front side of its body. And if the cat is already spinning, it can do all these motions to complement the existing rotation.

All cats are individuals, though. Long thin cats may employ a slightly different strategy to turn over than short fat cats, and individual cats might put a little more emphasis on one aspect of turning over as a matter of style or necessity. We have already seen, for example, that cats without tails can right themselves but that cats with tails will use those tails to speed up the motion. No two cats are exactly the same, and we should not expect any two cats to turn over in exactly the same way.

12

Falling Felines and Fundamental Physics

An explorer leaves his camp for a walk one morning. He walks one mile south, one mile east, and one mile north, which brings him right back home. Going back into his tent, he hears a commotion, looks out the flap, and sees a bear. What is the color of the bear?

Despite research stretching over three hundred years of physics history, cats have yet one more surprising secret about their ability to right themselves. The problem of cat-turning can be related to a concept known in physics as a *geometric phase*, a change in the condition of a system that is entirely due to the underlying geometry—real or mathematical—of the system itself. Through this connection, we can compare the falling cat to phenomena in quantum physics, in the behavior of light, and even in the motions of pendulums on the rotating Earth. Falling felines really do possess deep connections to fundamental physics.

To understand the concept of a geometric phase, it is helpful to start by thinking about motion on a familiar surface with a non-trivial geometry: our very own planet. The quotation at the beginning of the chapter is a variation on a classic brainteaser. There are two baffling but related aspects to the puzzle. Why doesn't the explorer walk one mile west to complete the circuit and arrive home? And what does the color of the bear have to do with any of this?

The answer: The bear is white. It is a polar bear. The tent must lie at the North Pole, one of two locations on the Earth where all the lines of longitude converge (the other location being the South Pole). Starting at the North Pole, a person walking south, east, and then north is following the path of a triangle, with the tent at the top vertex.

One lesson that can be taken away from this riddle is that the geometry of a sphere, such as the Earth, is surprisingly strange.[1] The lines of latitude and longitude that we use to define location on the Earth are almost everywhere perpendicular to each other; however, because these lines have been drawn on a sphere, there are two points where this description causes confusion, namely the North and South Poles. The circular longitude lines, which describe east-west location, all cross at the poles, and the circular latitude lines, which describe north-south location, shrink to points at the poles. The geometry of a sphere is fundamentally different from the geometry of a plane; any attempt to draw a plane on a spherical surface, or vice versa, will run into similar problems. This is why flat maps of the Earth use "projections" that inevitably distort the shapes and sizes of landmasses near the map edges. The famous Mercator projection, for instance, falsely makes Greenland look almost as large as the United States and makes Antarctica look as big as all of the continents combined—an artifact of stretching the top and bottom of the sphere to make a flat rectangular map.

A geometric phase is a change in the condition of a system that is due entirely to its being moved along a weird-shaped surface, such as a sphere. One example is a common fixture in many science museums: a massive free-hanging pendulum centered on a compass-like disk.

Foucault's pendulum on display at the London Polytechnic Institution in May 1851. From The World of Wonders.

This Foucault pendulum, named for its creator, Léon Foucault, was introduced to the public in 1851 to wide acclaim and has remained a topic of curiosity ever since. The reason for its popularity is that it shows simply and directly that the Earth is rotating. At first glance, the pendulum seems to be oscillating back and forth along a line passing through the center of its compass disk. Anyone watching for a few minutes, however, will see that the path of the pendulum slowly changes, turning one way or another around the disk like the minute hand of a watch.

But the pendulum itself is not changing direction. In fact, the Earth is rotating underneath the free-hanging pendulum. If a

Foucault pendulum were hanging at the North Pole, over the course of twenty-four hours the pendulum's direction of oscillation would appear to rotate a full 360 degrees as the Earth turns beneath it. It would return to its starting position by the end of the day. If the pendulum were hanging instead at the South Pole, it would appear to rotate in the opposite direction. So a Foucault pendulum is a direct and simple way to *see* the rotation of the Earth.

Léon Foucault, born in Paris in 1819, had never dreamed of being a scientist. Though he showed mechanical aptitude at a young age, his initial career aspiration was to enter the medical profession. He found himself unable to bear the sight of blood, however, which caused an abrupt change of career into science. He first worked as a lecturer's assistant, but his ingenuity and intelligence soon earned him acclaim as a researcher.

Foucault hit upon the idea of his pendulum while building astronomy equipment. He had placed a flexible steel rod into the end of a lathe, parallel to the axis of rotation of the lathe, and had inadvertently caused the rod to vibrate. Foucault noticed that the rod continued to vibrate along the same line, even as the lathe was turning; in a wonderful deductive leap, he realized that any freely vibrating object on the Earth must similarly vibrate independently of the Earth's rotation. A pendulum was a natural device with which to test this idea.[2]

Foucault first set up a small pendulum in a cellar, with a wire 2 meters (6 1/2 feet) long and a brass sphere weighing 5 kilograms (11 pounds). To make sure that the pendulum would swing in a straight line and not have any side-to-side or elliptical motion, he hung it off its center position with a piece of thread. After he burnt the thread, the pendulum would be released to swing freely. At the bottom of the pendulum sphere, Foucault affixed a small needle to scrape the ground so that he could observe minuscule changes in the direction of oscillation. In less than a minute, he noted that the pendulum direction had moved slightly but noticeably to the west, indicating that the Earth was rotating to the east.

The period of oscillation of a pendulum increases as the pendulum is made longer; a longer pendulum will show a larger displacement between swings than a shorter one. Also, a heavier pendulum is less likely to have its delicate motion disturbed by air currents or imperfections in its hanging mount. Foucault was well aware of this, so after his initial experiments at home, he set up a pendulum with a length of 11 meters (36 feet) at the Paris Observatory. In only two swings of this pendulum, the shift to the left could be clearly seen. Further emboldened, he set up his largest pendulum, with a length of 65 meters (213 feet), in the dome of the Pantheon in Paris. This pendulum became internationally famous, though it stayed in the Pantheon only until 1855. In 1995 a replica was installed in the original location, and it has been swinging ever since.

Foucault's experiment was a worldwide sensation. Crowds were drawn to the Pantheon to see the pendulum in action, and within a few short months, duplicate experiments were set up all over the globe. Audiences would sit and listen to a scientific lecture by a distinguished scientist. Afterward, they could see for themselves the change in the direction of the pendulum. According to a 1856 publication, "Throughout the world, a pendulum mania extended, until a monster pendulum threatened to become essential to every respectable household."[3]

Pendulum watching may seem an odd pastime, especially since most people, and all scientists, had already accepted the rotation of the Earth by the time of Foucault's discovery. But his pendulum allowed this motion to be seen in a clear and inarguable manner. It was a way for scientists to bring the motions of the cosmos into the lecture hall.

One peculiar aspect of the motion of the pendulum did not appear to trouble anyone in Foucault's time, but it would have profound consequences. If a pendulum is installed at the North Pole, it will rotate 360 degrees over the course of a day. A pendulum placed at the Equator, however, will not rotate at all. In both cases, the pendulum is oscillating along the same line at the end of the day as it was at the beginning. But what happens when a pendulum is placed at an intermediate latitude, such as at the Pantheon in Paris? Over the course of a day, the pendulum will rotate its plane of oscillation by an amount less than 360 degrees. At the Pantheon, the pendulum turns approximately 270 degrees during a full day.

This is strange. If we neglect the motion of the Earth around the Sun, which plays no significant role in Foucault's experiment, then after a day the pendulum could be said to have made a complete circular path around its latitude line and returned to its starting point in space. But the pendulum is now oscillating in a different direction! Somehow the path of the pendulum around the sphere of the Earth resulted in a different ending behavior for the pendulum from the one it started with.

To understand how this is possible, let's do a thought experiment. Imagine that you are carrying a small Foucault pendulum on a cafeteria tray. Suppose first that you walk in a circle, ending in the same location you started from. If you walk curving to the left the entire time, the direction of the pendulum will appear to turn to the right, and it will end up swinging along its original line once you return to your original position.[4] Of course, the pendulum is not changing direction during this process, you are—but the apparent direction of the pendulum changes because you are turning. Next, suppose you walk along a straight line. In that case, the pendulum will not change direction at all, but you will also not return to your starting location.

Now imagine that you are carrying the pendulum on a big spherical surface (not too hard to do, since we live on one). If you walk in a small circular path anywhere on the Earth, the pendulum will appear to swing 360 degrees again, just as it did when you walked on a flat surface, because a sphere is approximately flat over a small area of its surface. This is analogous to having a Foucault pendulum right at the North Pole, where the rotation of the Earth makes the pendulum appear to swing 360 degrees. You can also walk in a straight line on the surface of the Earth, though on a sphere a straight-line path is a great circle, such as the Equator or any circle that divides the sphere equally into two halves. The pendulum will not change direction as you walk on a great circle, but this straight-line path differs from the straight-line path on the flat surface because the shape of the sphere brings you back to your starting place, even though you never turned while you walked.

Finally, let us imagine that you walk with the pendulum along one of the northern latitude lines of the Earth. With the exception of the Equator, none of these latitude lines is a great circle; that is, they are not straight-line paths on the sphere. Therefore, if you walk with the pendulum along a latitude line that passes through, say, the Pantheon in Paris, you will constantly have to turn a little bit to the left to stay on that line. The direction of the pendulum will, consequently, appear to rotate to the right as you walk. However, because the shape of a sphere naturally has you heading back toward your starting point, you don't have to turn as much as you would on a flat surface to get back to that starting point. On a flat plane, you get back to your starting point by actively turning 360 degrees; on a sphere, you get back to your starting point partly by actively turning and partly by following the curvature of the Earth.

Foucault's pendulum therefore illustrates a geometric phase. That is, the underlying geometry of the Earth has allowed the pendulum to end up in the same place but not in the same condition it had at the start. Something very similar happens to the falling cat. It starts with its body upside down in an untwisted shape and proceeds to make a number of internal twisting and turning motions. After the cat makes those motions, its body returns to its original untwisted shape (the same "place"), but the cat is now right-side up (in a different "condition"). The twisting and turning of the cat is analogous to moving the pendulum around the Earth, and the change in the cat's orientation is analogous to the change in the pendulum's direction. Mathematically, a system that exhibits such a change is nonholonomic, or exhibits *anholonomy*.

Different types of anholonomy exist. For another example, let us return to our polar wanderer. We now ask, How does the elevation of the explorer change as he walks his path? He might climb a hill at some point along the path, increasing his elevation, but he will walk downhill at a later point in his path, so his altitude is the same again when he gets home.

Now suppose he walks inside a multilevel parking garage with spiral ramps connecting the levels. If the man's path takes him up one of the spiral ramps, he will be walking constantly uphill as he follows his closed loop and will end up exactly one floor above where he started. This is another example of an anholonomy: though the man has walked what is a closed path in terms of north-south-east-west coordinates, he has ended up at a new elevation. Similarly, the pendulum ends up in a different direction of oscillation, and the cat ends up with a different orientation.

Researchers in Foucault's time do not seem to have been particularly impressed by the anholonomy of his pendulum; they were no doubt more awed by the demonstration of the Earth's rotation and concerned with deriving the exact mathematical equations that describe its motion. It would be over a century before anholonomy in physics would be truly recognized and appreciated, in a very different context: quantum physics.

Physicists have accepted for nearly a century that everything in existence has a dual nature as a wave and a particle, a curious state of existence called wave-particle duality and which, we will see, resulted in the conception of Schrödinger's cat. When a single quantum particle, such as an electron, is confined to a closed region, its wave properties result in the existence of certain stable and relatively simple motions. These states of motion are paradoxically known as *stationary states*, and each has a well-defined discrete energy associated with it. We can visualize this by considering a vibrating string, which mathematically is similar to a quantum particle being trapped in a one-dimensional box. Though the string can be vibrated at any frequency (energy), certain frequencies make the string vibrate in a very simple manner; these are the stationary states of the string. These states can be demonstrated with a thick rope, such as a jump rope or old coiled phone cord, tied to a solid object at one end and held slightly taut. Rapidly shaking it creates "natural" modes similar to those in the accompanying figure.

Quantum particles, or vibrating waves, can be excited in boxes of more complicated shapes. For example, waves will vibrate on a circular drumhead, which is analogous to a quantum particle being trapped in a circular box; stationary states will be associated with that drumhead. For simple shapes, such as circular boxes or rectangular boxes, we can mathematically solve for the energies of the stationary

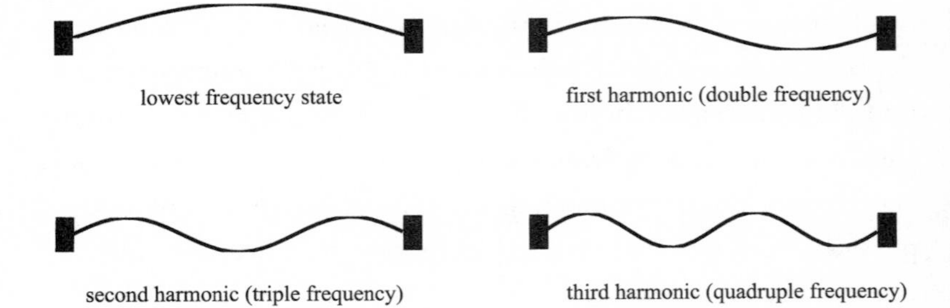

The lowest-frequency stationary states of a vibrating string. For a quantum particle, the energy is proportional to the frequency. My drawing.

states; these basic calculations are taught in an undergraduate physics education.

For boxes with more complicated shapes, however, the calculations often cannot be done directly; finding the stationary states can be quite difficult. In the late 1970s, Michael Berry of the University of Bristol wanted to understand the stationary states in such cases. In particular, he was concerned with finding systems in which two or more of the stationary states end up having the same energy, a situation referred to as a *degeneracy*. Such degeneracies are, in the problems Berry was investigating, infinitely rare; the only way to find them is to mathematically study an entire class of systems simultaneously and isolate those systems where a degeneracy occurs. Similarly, anyone hunting for a four-leaf clover must search an entire field of clover to spot a four-leaf specimen instead of the much more common three-leaf variety.

The problem that Berry eventually investigated was the case of a quantum particle bouncing within a triangular box, which is analogous to waves vibrating on a triangular drumhead.[5] By studying the stationary states that arise for every imaginable triangular shaped box, it would be possible to find those boxes with degeneracies. In the context of this problem, these degeneracies are named diabolical points because of their relationship to a double-cone structure, a diabolo (and not because of any inherent wickedness).

The shape of a triangle can be characterized by two parameters—namely, two of its interior angles, which we label X and Y. Because the three angles of every triangle add up to 180 degrees, the angle of the third side is fixed by the choice of the other two. So Berry and his colleague Mark Wilkinson devised a mathematical technique to hunt in boxes for diabolical points with every possible value of X and Y. But how would they know when a diabolical point had been found? Here, they found a curious property of the system in question. Since a diabolical point involves two distinct stationary states within a triangle that have exactly the same energy, Berry and Wilkinson found that if they mathematically "walked" around in their collection of triangles, treating X and Y as the latitude and the longitude of the path walked, the waves of the two stationary states would flip upside down over the course of the walk if the path contained a diabolical point.

Here, we can draw a direct analogy with our multilevel parking garage. Just as the polar explorer will end up on a different level of the garage if he walks up a ramp of the garage, the waves of the triangle will flip sign if one "walks" around a diabolical point—the "up" part of each wave will become "down," and vice versa. A key difference is that the walk in the parking garage is a walk in real space, while the walk of Berry and Wilkinson is a theoretical walk through a mathematical construct. Using this technique, they found a number of diabolical points in their set of triangles.

The change of the wave in this case should properly be called a *topological phase*. Topology is the field of mathematics that distinguishes objects by the way their constituent parts are connected; a sphere, for example, is different from a donut, for a donut has a hole in it, and a sphere does not. The parking deck, in turn, is different from a set of stacked parallel planes because of the ramp connections. With Berry and Wilkinson's topological phase, the most change they could expect when going from one "level" to another was for the wave to flip sign.

This topological phase hinted at a profound breakthrough to come. The "moment of conception," as Berry put it, came in the spring of 1983, when he presented his work at the Georgia Institute of Technology. Berry had already noted that the diabolical points could exist only if magnetic fields were not influencing the particle in the triangular box. He continued,

> So, if a weak magnetic field were added to the particle in the triangles, the diabolical points would disappear. At the end of the talk, Ronald Fox (at that time the chairman of the physics department) asked what happens to the sign change when the magnetic field is switched on.
>
> This was the trigger, the moment of conception. My immediate response, "I suppose it's a phase change different from π," was followed by the premature "I'll work it out tonight and tell you tomorrow." In fact it took several weeks to understand the geometric phase properly.[6]

Berry had recognized that a quantum particle, taken through a slow sequence of changes and brought back to its original condition, can nevertheless end up in a different state from the one it started with. He further showed that the change that accumulates in this process depends on the underlying mathematical geometry of the quantum system in question—that is, it is a *geometric phase*. Berry had stumbled across a previously unappreciated general feature of many quantum systems. He published his work in 1984.[7]

In the case of the triangular box, suppose that we start with an electron in a box shaped like an equilateral triangle and apply a magnetic field. Then the box shape is slowly distorted in such a way that the total change in X and Y takes it continuously through different shapes and then back to being an equilateral triangle. Berry showed that, although the box has ended in the same shape as it began, the wave of the quantum particle will have accumulated a different phase—one that has to do with the mathematical geometry of the complete set of triangular boxes.

Here, the connection to the Foucault pendulum becomes explicit. Just as a pendulum, brought in a closed path around a latitude line on the Earth, ends up oscillating in a different direction from the one it started in, a quantum particle, brought in a closed path through some set of system parameter changes, ends up in a different state from the one it started in. A cat, too, making a sequence of motions with its body segments and returning its body to its original shape, ends up with a different orientation from the one it started with. Both the falling cat problem and the Foucault pendulum represent examples of this accumulated change of state called a geometric phase.

In spite of the groundbreaking nature of the work, Berry was initially discouraged when he learned that other researchers had made small steps in the same direction. One pair of authors had noted similar phases in the waves of colliding atomic nuclei in 1979.[8] However, Berry's situation was somewhat like Einstein's. After Einstein published his 1905 paper on the formulation of the special theory of relativity, it was recognized that numerous other researchers had been putting together isolated bits and pieces of the theory; it was only Einstein, however, who compiled these observations and showed their broader significance. Similarly, Berry's recognition of the geometric phase demonstrated that it touches on countless aspects of quantum physics and beyond.

One area to which the theory had already unknowingly been extended was optics. While Berry was visiting India in 1986, his colleagues drew his attention to work on the polarization of light done in the 1950s by Shivaramakrishnan Pancharatnam.[9] When James Clerk Maxwell demonstrated that light is an electromagnetic wave back in the 1860s, he simultaneously demonstrated that light consists of an electric wave and a magnetic wave oscillating together and perpendicular to the direction the wave is traveling; the manner in which the electric wave is oscillating is referred to as the *polarization* of light. If we could look straight into a beam of light and see the rapid oscillations of the electric wave, its path would appear very much like one of the possible motions of a Foucault pendulum when viewed from above.[10]

The "state" of polarization is the shape of the ellipse formed by the electric wave and the angle of the ellipse, and it can be changed by various optical devices (such as polarized sunglasses). Pancharatnam investigated the behavior of light as the polarization is changed continuously from a starting state through a variety of different states and back to the original one. He found that the oscillations of the electric field ended up slightly out of sync with those of the original polarization state, an effect that could only be attributed to the particular way that the polarization had been changed. Pancharatnam had found an early example of a geometric phase. Not long after this connection was recognized, Berry wrote a paper explaining the relationship, with due credit to Pancharatnam.[11]

The connection between falling cats and the geometric phase took a bit more time to be made. In 1990, Jerrold Marsden, Richard

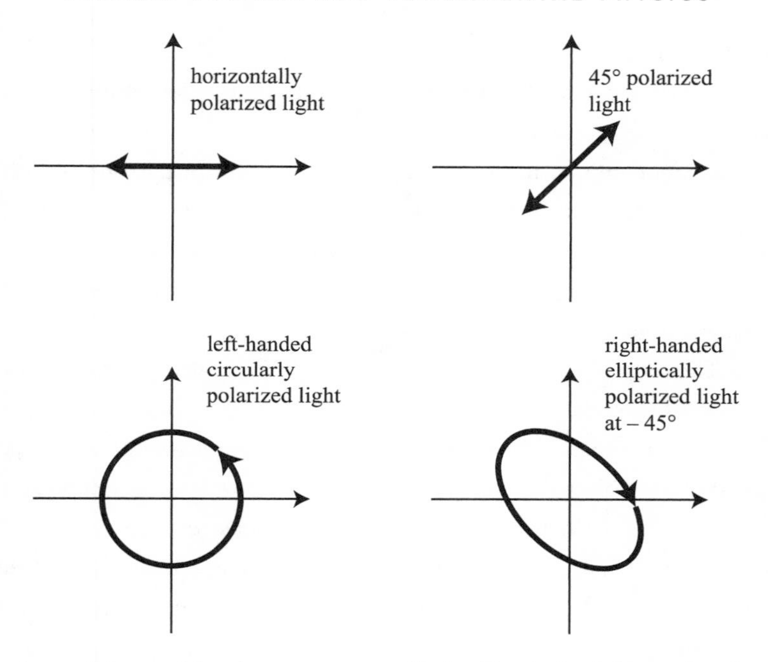

Some of the possible polarization states of light. The thick solid lines show how the electric wave oscillates as a person looks head-on into the beam of light. My drawing.

Montgomery, and Tudor Ratiu wrote an extended monograph on the implications and applications of geometric phases to problems of mechanical systems with lots of moving parts; the cat is given a brief mention in this context: "In this setting one can formulate interesting questions of optimal control such as 'when a cat falls and turns itself over in mid flight (all the time with zero angular momentum!) does it do so with optimal efficiency in terms of say energy expended?' "[12] As an example of turns with zero angular momentum they give a human version of what we can recognize as Peano's propeller-tail model of the falling cat, which they referred to as "Elroy's beanie." Let us imagine a person, in an upright position, wearing a beanie hat with a propeller on the top. If the person is in freefall, and the propeller on the beanie spins, the person must spin in the opposite direction because of conservation of angular momentum. However, since the person is much heavier than the propeller, the person's body will rotate only a slight amount every time the propeller turns. After the propeller returns to its starting position, the overall person-beanie combination will have a slightly different orientation.

The most thorough discussion of the falling cat in terms of geometric phases appeared in 2003, written by the physics philosopher Robert Batterman.[13] In it, he ties together falling cats, the Foucault pendulum, polarized light, and even parallel parking as manifestations of the geometric phase in physics; the last example is worth explaining briefly. In parallel parking, the car is effectively moved sideways through the use of forward and backward motion and turns. The "phase" in this case is the sideways position of the car, which has changed even though the orientation of the car is the same at the beginning and end of the maneuver.

One important lesson in the discovery of geometric phases is that many complicated physics problems have a beautiful underlying geometry. In the case of the Foucault pendulum, the geometry is a real one—the spherical shape of the Earth—but with falling cats, quantum particles, and light polarization, similar geometries can be found hiding in the mathematics of the problem. Once this geometry has been revealed, the problem becomes much easier to understand, almost trivial in some cases.

For the Foucault pendulum, let us imagine that we evaluate the behavior of the pendulum by constructing a model of the Earth with a radius of one. We then trace out the path of the pendulum on the surface of the sphere. We can show mathematically that

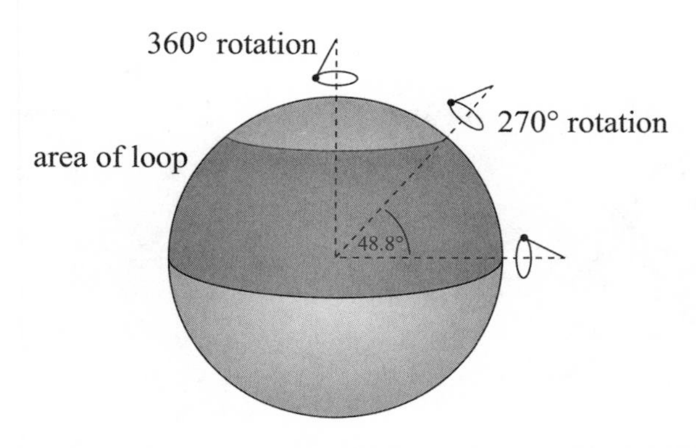

The amount of angle by which the pendulum rotates over twenty-four hours is equal to the surface area of the strip carved out between the Equator and the line of latitude at which the pendulum swings. My drawing.

the angle by which the pendulum rotates at the end of twenty-four hours—measured in radians, not degrees—is equal to the surface area of the surface carved out between the Equator and the latitude line.[14]

For the geometric phase discovered by Pancharatnam, we can determine the lag created in the light wave by using a *Poincaré sphere*. It can be shown that every possible state of light polarization can be mapped to a point on a sphere with a radius of one unit. On the Poincaré sphere, the North and South Poles of the sphere have left-handed (counterclockwise) and right-handed (clockwise) circular polarization; the Equator has every state of linear polarization; and the Northern and Southern Hemispheres represent every possible state of left-handed and right-handed elliptical polarization, respectively. Any continuous change in the polarization state of light can be drawn as a path on the Poincaré sphere, like the path that the GPS in a car traces out from starting to ending destinations. If the path is closed—that is, if the polarization is brought back to its original

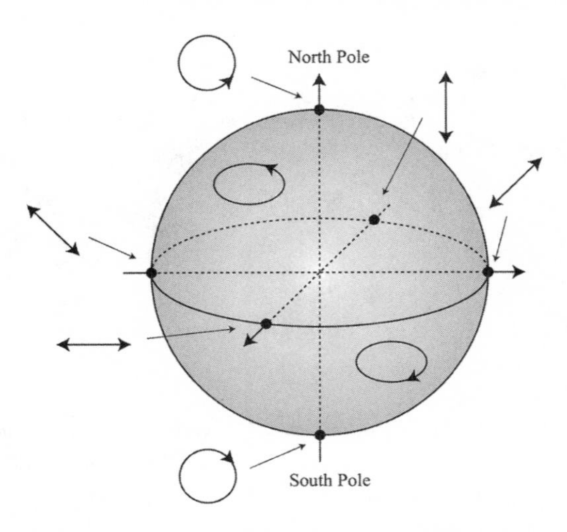

The Poincaré sphere. Every possible state of light polarization can be located on the sphere, and any cyclic change in the state of polarization can be drawn as a closed path on the sphere. My drawing.

state—the Pancharatnam phase accumulated by the light during the change is given by *half* the surface area that the path covers on the sphere.

The cat problem can also be connected to the surface area of an appropriate geometric shape. For a cat with a very specific length-to-girth ratio, we can describe the orientation of the cat—its geometric phase—with a sphere.[15] Using Rademaker and ter Braak's bend-and-twist model, we can say that the latitude on the sphere represents the amount the cat bends at the waist, while the longitude represents the amount it twists at the waist; see the nearby illustration. Any bending and twisting of the cat can therefore be modeled as a path on the sphere, and it can be shown that the overall rotation of the cat in righting itself is equal to the surface area that the path makes on this "cat sphere."

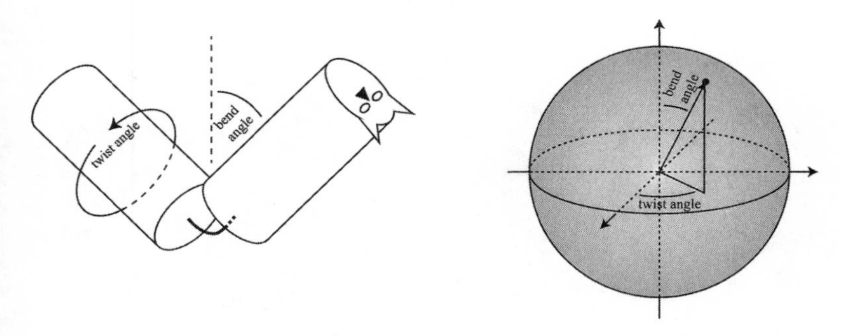

Rademaker and ter Braak's cat model together with the "cat sphere" that can be used to calculate the amount of the cat's overall rotation. My drawing.

The true beauty of the geometric phase, then, is that it allows us to solve very complicated problems through the use of very simple geometry. Let us look again at the sophisticated cat-turning model of Kane and Scher, developed in the late 1960s for NASA. A big limitation of the earlier model of Rademaker and ter Braak was the assumption that a cat keeps the same bend in its spine through its entire twist, even though a cat certainly can't bend backward as well as it can forward. Kane and Scher's model results in a cat that decreases its back bend as it twists, and effectively flips from one side to the other at the height of its twist.

If we compare both motions on the cat sphere, we can see the limitations of Rademaker and ter Braak's model as well as the sensibility of Kane and Scher's. Looking at the cat sphere from above, we can see that in the simple Rademaker and ter Braak model the cat would have to perform an extreme back bend. The Kane and Scher model, by contrast, avoids that extreme back bend. There the cat switches from a right bend to a left bend before completing the motion.

Once we look at the cat sphere from above, we can readily see that the Kane and Scher model is the optimal choice: it maximizes

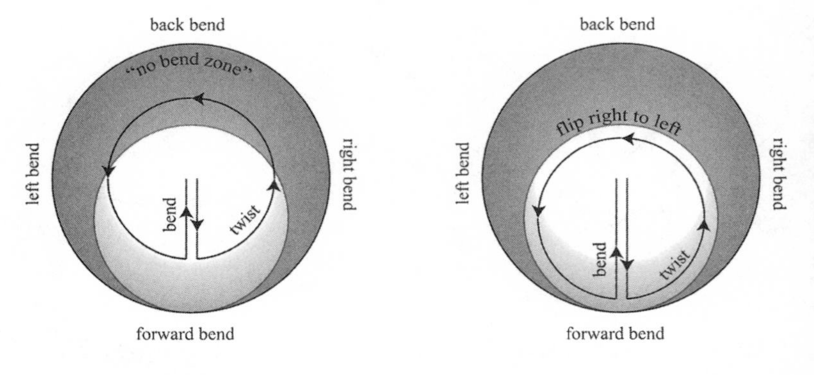

The cat sphere viewed from above, comparing Rademaker and ter Braak's model with Kane and Scher's model. My drawing.

the amount of surface area enclosed in the path—and therefore maximizes cat twist—without requiring the cat to bend in an impossible way. From an evolutionary perspective, the cat's motion has been honed to take advantage of all the bending and twisting available.

The description of cat-turning through the use of a spherical surface brings us, ironically, full circle from the original work of Antoine Parent over three hundred years ago. Parent, out of mathematical convenience, modeled a falling cat as a sphere. Now, looking at the phenomenon of cat-turning as a geometric phase, we see that a cat can, in fact, be modeled as a sphere, albeit in a very different manner than Parent ever envisioned.

The geometric phase is, perhaps, the last profound secret that the falling cat has been keeping. Though the geometric phase was first recognized as a general phenomenon by Berry in the 1980s, it baffled physicists back in 1894, when Marey showed his photographs of a falling cat to the Paris Academy. It took about one hundred years for the falling cat problem to be recognized as a geometric phase phenomenon: cats can hide their secrets very well.

But is the geometric phase connection the last secret that cats are hiding? Researchers keep finding more connections between falling cats and subtle physical problems. In 1993, Richard Montgomery wrote a paper on the "gauge theory of the falling cat" using very sophisticated mathematics to describe cat-turning. This work was followed in 1999 by Toshihiro Iwai, who considered the problem of zero-angular-momentum turns in the context of quantum physics; in his article he gave due acknowledgment to the falling cat.[16]

But the most mind-bending paper was one written in 2015 by researchers in Mexico, with the provocative title, "Do Free-Falling Quantum Cats Land on Their Feet?"[17] In it, the researchers consider a purely quantum-mechanical cat falling and find that it can land in a Schrödinger-cat state—that is, simultaneously right-side up and upside down.

Cats have long been said to possess nine lives. The physics of falling cats may have one life left in it as well.

13

Scientists and Their Cats

Throughout this book, we have seen countless examples of cats being used as little more than test subjects. As an antidote to this rather dark view of scientific progress, it is worth recognizing that many physicists throughout history have had a much more amicable relationship with their feline acquaintances, employing them as laboratory assistants, muses, companions, and even scientific co-authors. We conclude with a look at some of these fascinating collaborations.

To start off, let me debunk the story that Isaac Newton not only made world-changing discoveries involving motion and gravity but also invented the cat door.

Newton himself could be considered catlike in his personality, for he was clever, mischievous, somewhat solitary in his pursuits, and fierce. In a list of confessed sins that he wrote down at the age of nineteen, Newton included the sin of "Threatening my father and mother Smith to burn them and the house over them."[1] Newton was not on good terms with his stepfather, the Reverend Barnabas Smith, and was evidently angry at his mother for remarrying.

The story of Newton's cat door was regularly repeated in the 1800s. One colorful version is presented below in full.

> You know the anecdote of Coleridge, Southey, and Wordsworth, whose united wits could not get the horse's collar off over

> his head, when the Scotch girl did it with a single jerk. The Country Parson tells a much better anecdote than that, of Sir Isaac Newton, which, by the way, we never saw anywhere else. The great philosopher had a pet cat and kitten, which he harbored in his study; but becoming tired of opening the door for them to go out and in, he hit upon the following contrivance. "He cut in his door a large hole for the cat to go out in, and a small hole for the kitten. He failed to remember what the stupidest bumpkin would have remembered, that the large hole through which the cat passed might be made use of by the kitten too. Having provided the holes, he waited with pride to see the creatures pass through them for the first time. As they arose from the rug before the fire where they had been lying, the great mind stopped in some sublime calculation; the pen was laid down; and all but the greatest man watched them intently. They approached the door, and discovered the provision made for their comfort. The cat went through the door by the large hole provided for her, and instantly the kitten followed her THROUGH THE SAME HOLE." As ministers are constantly charged with the same want of common sense, it may be consoling to find ourselves in company with the poets and philosophers. It is, however, a positive loss, not only of convenience, but of power and influence, if one fails to be developed on the practical side.[2]

Here the story of Newton's cat door is used as a lesson on humility, directed especially at those hubris-filled philosophers. "Sure, you can predict the motion of the planets, but that knowledge is useless without some old-fashioned common sense!"

But is the cat door story true? Newton certainly did not invent the idea of a cat door, which existed in some form or other for hundreds if not thousands of years before Newton. Geoffrey Chaucer's *Canterbury Tales* (1386), for instance, contains a passage involving a "cat hole," as such portals were called. In "The Miller's Tale," a servant knocking at an exterior door of a home, receiving no response, uses a cat hole to peek inside.

> An hole he foond, ful lowe upon a bord
> Ther as the cat was wont in for to crepe,
> And at the hole he looked in ful depe,
> And at the last he hadde of hym a sighte.

Furthermore, did Newton even have a cat? There is no evidence that Newton had a pet at all—cat, dog, or otherwise. None is mentioned in his correspondence or that of colleagues. Though Newton's family home, Woolsthorpe Manor, is still in existence, there is no evidence of a cat door or doors. That is not proof either way, though. The house doors have probably been replaced in the hundreds of years since Newton's time.

The apparent origin of Newton's cat door story comes from mathematician John M. F. Wright, who worked at Trinity College and published a memoir of his experience there in 1827.[3] He heard many tales of former Trinity scholar Newton and shared a number of them. Of Newton and animals, he had the following to say.

> Many anecdotes are told of Newton's "absence of mind," ascribable to this total abstraction from the world and its ways, which, to such readers as may not have heard them, will doubtless prove amusing.
> . . .

In all the operations of nature, it is her simplicity which strikes the intelligent beholder. The Son of God was simple, meek, and lowly, as a lamb. Newton, who is allowed to have risen highest of those creatures whom God "formed after his own image," was also the simplest of mankind. It is said that this great practical, as well as theoretical philosopher, having on a certain occasion, left a candle burning near to a pile of papers, the result of years of fatiguing study, and doubtless like all his other writings, full of brilliant discoveries, his dog Dido upset it, so that they were utterly destroyed. The only lamentation he sent forth on his return, was, "Dido, Dido, thou little knowest the mischief thou hast done."

On another occasion, a friend, for the joke's sake, having eat the chicken served up for his dinner, on seeing the bones, Newton exclaimed, "How forgetful I am! I thought I had not dined," and returned to his meditations, supposing all was as it should be. Being much attached to domestic animals (although for woman, sweet woman, he seems to have had no impulses absolutely irresistible, inasmuch as he is said always to have resisted the tender passion—or, at all events, to have so highly valued the daughters of Eve, as to have deemed them worthy of all or nothing of a man's time and attention), he possessed himself of a cat as a companion to his dog Dido. This cat, in the natural course of events, although Newton had probably not calculated upon it, produced a kitten; [then] the good man, seeing at a glance, the consequences of the increase of his family, issued orders to the college carpenter to make two holes in the door, one for the cat and the other for the kitten. Whether this account be true or false, indisputably

> true is it that there are in the door to this day two plugged holes of proper dimensions for the respective egresses of cat and kitten.

These stories, seemingly handed down through several generations of students and scholars at Trinity, are undoubtedly embellished or even completely made up. Recall the story of Maxwell throwing cats out of windows, which had evolved dramatically only twenty years after Maxwell left. Even if the holes were cat doors, there is no guarantee that Newton put them there. It seems much more likely that a student saw the holes in the door and let his imagination run wild.

It is easy, however, to imagine a scientist like Newton, working alone in a laboratory for hours on end, relying on a cat for company. Indeed, another key figure in the physics of motion had cat companions to help him in his studies. William Rowan Hamilton (1805–1865) was an Irish astronomer, physicist, and mathematician whose most famous work is a sophisticated mathematical reformulation of Newton's laws of motion. Hamilton's laws of motion turned out to be ideally suited to the analysis of quantum particles, and a function known as the *Hamiltonian* finds use in both classical and quantum physics.

Hamilton was renowned for his kindness to both humans and animals—cats in particular—as his sister attested.

> He was always fond of cats, and might often be seen writing some mathematical paper with a kitten or favourite cat on his shoulder playfully trying to catch the pen.
>
> His politeness was almost a rebuke to others. A young lady, who was living with me in Dublin at one time said, "I never

saw so polite a gentleman as your brother; I think he would almost bow to a cat"; and I was reminded of her and amused him by repeating this to him one day, when accidentally he did tread upon the cat's paw, and turned round, and smiling said, "I was going to say, I beg your pardon."

But it was not just his personal animal companions that he treated with such patience. As his sister recounted,

His feeling consideration for all living things around him drew towards him from them unbounded confidence. One instance of this made a deep impression on those who witnessed it, and indeed it was an occurrence fitted to excite and to excuse a some what superstitious wonder. On a Whitsunday morning, as he was reading prayers in the centre of his assembled household, a dove flew in through the open window and settled on his head; it was undisturbed by Hamilton, who continued to read, and after an interval it peacefully flew out.[4]

In addition to keeping cats as companions, other scientists, such as Robert Williams Wood (1868–1955), found ingenious ways for cats to help in experiments. Wood was a pioneer in the study of ultraviolet light and did extensive work in the field of spectroscopy, the study of the structure of matter through the measurement of the spectrum (colors) of light. He was a dedicated researcher and even did work during vacation time with his family, as described in a biographical memoir.

Wood and his family spent the summers on an old farm on Long Island. In a barn he had improvised a laboratory, and

> one of its features was this forty-foot grating spectrograph, probably the largest then in existence and certainly capable of giving better results than anyone had ever seen before. It was constructed from sewer pipe laid by the local stonemason. During the long months between summers when the instrument was not used, all sorts of wildlife used it as a shelter, and the optical path became cluttered up with spider webs. Wood's method of cleaning the tube has become a classic. He put the family cat in one end and closed this end so that the cat, in order to escape, had to run through the whole length of the tube, ridding it very effectively of all spider webs.[5]

The cat may not have been a willing lab assistant, but it does not seem to have come to any harm, either.

Wood's imagination was not restricted to finding novel uses for a feline. He also co-authored two science fiction novels, *The Man Who Rocked the Earth* (1915) and *The Moon-Maker* (1916), both with Arthur Train. The former concerns a mysterious scientist who forces the warring countries of the world into a lasting peace through the threat of a nuclear blast that can alter the Earth's rotation; here, Wood and Train were somewhat presciently describing the eventual creation of nuclear weapons (though the details are obviously wrong). In *The Moon-Maker*, a sequel, scientists travel into space to deflect or destroy an asteroid that is on course to annihilate the Earth, anticipating the plot of the 1998 movie *Armageddon* by eighty-two years.

Though his scientific discoveries are significant, Wood is perhaps best known for disproving, in a simple and devastating way, the existence of "N-rays" in the early twentieth century. In the era when X-rays and "uranic rays" (radioactivity) had been discovered, it seemed that new forms of invisible radiation were everywhere. Working in Nancy, France, Prosper-Ren Blondlot thought he had discovered a new class of rays that, unlike X-rays, interacted with electricity. He named these new rays *N-rays* after his city of residence. Hundreds of research articles were published by a variety of scientists confirming Blondlot's results.

However, a far greater number of researchers could find no evidence of N-rays. Finally, in 1904, Wood traveled to Nancy to work with Blondlot to get to the root of the mystery. Noticing that the only evidence for the existence of N-rays was found in supposed changes in the wild flickering of an electrical spark, Wood removed a key piece of equipment from the experimental apparatus while nobody was looking. The researchers happily continued, not seeing any change in their results. Wood had shown that N-rays did not exist except in the overly optimistic imaginations of the French scientists.

Wood's science fiction writing, cat re-purposing, and N-ray debunking illustrate a childlike cleverness and sense of wonder about the world; in 1941 his biographer William Seabrook referred to him as "a small boy who never grew up." This was meant as the highest compliment.[6]

Some physicists have found inspiration and, indeed, a life's calling, in interactions with cats. The most notable example is Nikola Tesla (1856–1943), inventor, physicist, and futurist. Tesla is popularly known as the "Master of Lightning" for his work on electrical power generation; he developed and championed the alternating current (AC) power system we still use to this day. This work, which was sponsored by Westinghouse Electric and Manufacturing Company, led Tesla and Westinghouse into a business war with Thomas Edison, who was pushing direct current (DC) power in the United

Nikola Tesla in his Colorado Springs laboratory around 1899, calmly working while lightning crashes around him. The famous photograph is a double exposure made by Dickensen V. Alley. Wikimedia Commons, Wellcome Images.

States. Tesla also observed X-rays in 1894, beating Wilhelm Röntgen by a year, but lost his research records in a fire in March 1895. Tesla experimented with radio and the wireless transmission of electrical power, inventing the Tesla coil, a device that emits massive sparks and can make fluorescent light bulbs glow without being plugged in.

By all accounts, Tesla showed signs of brilliance even at a very young age. But what inspired him to study electrical phenomena, out of all the possible subjects that he could have excelled at was, in his own words, a cat.

In 1939, Tesla wrote a letter to Pola Fotich, the young daughter of the Yugoslavian ambassador to the United States.[7] In it, he describes his childhood home in Yugoslavia and talks about his feline companion.

> But I was the happiest of all, the fountain of my enjoyment being our magnificent Macak—the finest of all cats in the world. I wish I could give you an adequate idea of the affection that existed between us. We lived for one another. Wherever I went, Macak followed, because of our mutual love and the desire to protect me. When such a necessity presented itself he would rise to twice his normal height, buckle his back, and with his tail as rigid as a metal bar and whiskers like steel wires, he would give vent to his rage with explosive puffs: Pfftt! Pfftt! It was a terrifying sight, and whoever had provoked him, human or animal, would beat a hasty retreat.
>
> Every evening we would run from the house along the church wall and he would rush after me and grab me by the trousers. He tried hard to make me believe that he would bite, but the instant his needle-sharp incisors penetrated the clothing, the pressure ceased and their contact with my skin was gentle and tender as a butterfly alighting on a petal. He liked best to roll on the grass with me. While we were doing this he bit and clawed and purred in rapturous pleasure. He fascinated me so completely that I too bit and clawed and purred. We could not stop, but rolled and rolled in a delirium of delight. We indulged in this enchanting sport day by day except in rainy weather.
>
> In respect to water, Macak was very fastidious. He would jump six feet to avoid wetting his paws. On such days we went into the house and selected a nice cozy place to play. Macak was scrupulously clean, had no fleas or bugs, shed no hair, and

> showed no objectionable traits. He was touchingly delicate in signifying his wish to be let out at night, and scratched the door gently for readmittance.

So far, we have a simple story of a child's love for his pet. But the story takes a distinctly scientific turn.

> Now I must tell you a strange and unforgettable experience that stayed with me all my life. Our home was about eighteen hundred feet above sea level, and as a rule we had dry weather in the winter. But sometimes a warm wind from the Adriatic would blow persistently for a long time, melting the snow, flooding the land, and causing great loss of property and life. We would witness the terrifying spectacle of a mighty, seething river carrying wreckage and tearing down everything moveable in its way. I often visualize the events of my youth, and when I think of this scene the sound of the waters fills my ears and I see, as vividly as then, the tumultuous flow and the mad dance of the wreckage. But my recollections of winter, with its dry cold and immaculate white snow, are always agreeable.
>
> It happened that one day the cold was drier than ever before. People walking in the snow left a luminous trail behind them, and a snowball thrown against an obstacle gave a flare of light like a loaf of sugar cut with a knife. In the dusk of the evening, as I stroked Macak's back, I saw a miracle that made me speechless with amazement. Macak's back was a sheet of light and my hand produced a shower of sparks loud enough to be heard all over the house.

Young Tesla was noticing, for the first time, the phenomenon of static electricity. For many children, including myself, static electricity is the first introduction to the strangeness of the physical world.

I had an analogous introduction to the phenomenon, albeit a much more painful one. When I was perhaps six years old, my grandmother gave me a pair of wool slippers for Christmas. Walking on the carpet in wool slippers works very much like rubbing a cat's fur in generating static electricity. One day I started chasing my sister around the house after she had teased me. The kitchen, dining room, and living room formed a continuous loop, and I pursued her around and around. Our Christmas tree was in the living room on the inside of this loop. After about four orbits around the tree, I had built up enough static electricity that the metal tinsel on the Christmas tree reached out and shocked me, making me drop to the floor. My sister began laughing at me, and this caused me to get up and chase her again, leading in four revolutions to another electric shock, again and again.

Back to Tesla's account.

> My father was a very learned man; he had an answer for every question. But this phenomenon was new even to him. "Well," he finally remarked, "this is nothing but electricity, the same thing you see through the trees in a storm."
>
> My mother seemed charmed. "Stop playing with this cat," she said. "He might start a fire." But I was thinking abstractedly. Is nature a gigantic cat? If so, who strokes its back? It can only be God, I concluded. Here I was, only three years old and already philosophizing.
>
> However stupefying the first observation, something still more wonderful was to come. It was getting darker, and soon

the candles were lighted. Macak took a few steps through the room. He shook his paws as though he were treading on wet ground. I looked at him attentively. Did I see something or was it an illusion? I strained my eyes and perceived distinctly that his body was surrounded by a halo like the aureole of a saint!

I cannot exaggerate the effect of this marvelous night on my childish imagination. Day after day I have asked myself "what is electricity?" and found no answer. Eighty years have gone by since that time and I still ask the same question, unable to answer it. Some pseudo-scientist, of whom there are only too many, may tell you that he can, but do not believe him. If any of them know what it is, I would also know, and my chances are better than any of them, for my laboratory work and practical experience are more extensive, and my life covers three generations of scientific research.

Tesla's guiding question is strikingly similar to Albert Einstein's in 1951, when he looked back over fifty years of thinking about "light quanta" and realized that he still didn't know what they were. As he wrote to a friend, people who thought they knew were fooling themselves.[8] The light quanta that Einstein wrote about are what we now refer to as photons, discrete particles of light. As we have noted, Einstein introduced the concept of photons in his 1905 explanation of the photoelectric effect, which won him the Nobel Prize in Physics in 1921.

The statements by Einstein and Tesla highlight an important point about the philosophy of physics: physics can be very good at explaining *how* things work, through formulas and observations, but it doesn't necessarily tell *why* things work the way they do. Both Tesla and Einstein recognized that there were deep questions raised by their research that they had not even come close to grasping.

Einstein himself was fond of animals and at one point had a cat named Tiger who would get depressed when it rained. Einstein reportedly said to the cat, "I know what's wrong, dear fellow, but I don't know how to turn it off." In 1924, Einstein wrote a letter to friends, possibly about the same cat: "I have the urge to send you also a direct greeting from my hermitage, for once. It's so nice here that I could almost envy myself for it. I'm inhabiting an entire floor by myself. Nobody besides me is here other than sometimes a huge tom cat, who also mainly determines the smell in my room, as I am not able to compete successfully with him in that respect."[9]

Einstein had a much more powerful fascination with a cat of a different sort, one that never existed but is the most famous cat in all of physics: *Schrödinger's cat.* This strange beast was introduced by Austrian physicist Erwin Schrödinger (1887–1961) to highlight the seemingly absurd implications of theoretical quantum physics, a theory that Schrödinger himself had a large role in developing.

If we were to summarize the whole of quantum physics in a few words, we would say it declares that everything in existence has both a wave nature and a particle nature simultaneously: this, as we saw earlier, is referred to as wave-particle duality. In 1905, Einstein successfully argued that light, thought to be a wave since the early 1800s, also behaved as a stream of particles. In 1924 the French physicist Louis de Broglie, inspired by Einstein's observation, suggested that the reverse is true for matter: that all atoms and all elementary particles, such as electrons, have wavelike properties. This hypothesis by de Broglie was confirmed experimentally only a few years later, truly ushering in the era of quantum physics. In 1926, Erwin Schrödinger created a mathematical formula—the *Schrödinger equation*—that described how the wave properties of matter evolve in time and space. Einstein endorsed his work for publication.

There was a very big problem, however. Everyone agreed that matter behaves as a wave, but nobody could explain, precisely, what is waving. When we speak of water waves, we are talking about the water itself moving up and down; when we speak of sound waves, we know that it is the air molecules that oscillate, carrying the sound from source to receiver. For light, as James Clerk Maxwell recognized, the electric and magnetic fields are doing the "waving." But nobody was exactly sure how to think about waves of matter. As Tesla and Einstein said about electricity and photons, respectively, it is one thing to describe a phenomenon, quite another to interpret it.

This question became the obsession of the Danish physicist Niels Bohr and the German physicist Werner Heisenberg. Heisenberg worked under Bohr at an institute in Copenhagen. Together they compiled all existing knowledge of quantum physics into a coherent system that is known today as the *Copenhagen interpretation* of quantum mechanics. Briefly: the matter wave of electrons and other particles is not a physical wave but is instead related to the probability that a particle will appear in a particular location at a particular time. A tall wave represents a high probability that the particle will appear, while a low wave represents a low probability.

When we measure the location of a quantum particle, however, we never see a spread-out wave: we see a particle localized to a point in space. A key component of the Copenhagen interpretation is the idea of *wavefunction collapse*: when anyone tries to measure the position of a particle, its wave "collapses" to a single location, which represents the location of the particle. By implication any measurement of a quantum particle dramatically changes its behavior. Furthermore, the Copenhagen interpretation suggests that the quantum particle, before it is measured, is not definitely in one place or another. Only when it is measured does it "decide," in some mysterious manner, where exactly it wants to be.

If this interpretation is unsatisfying, you are in the company of Schrödinger himself, who concluded that the Copenhagen interpretation could lead to absurd results. He noted that it is possible to make a living creature, such as a cat, have its very existence dependent on the behavior of a single quantum particle, such as an atom. "One can even set up quite ridiculous cases," he wrote in 1935, and gave an example.

> A cat is penned up in a steel chamber, along with the following diabolical device (which must be secured against direct interference by the cat): in a Geiger counter there is a tiny bit of radioactive substance, so small, that perhaps in the course of one hour one of the atoms decays, but also, with equal probability, perhaps none [decay]; if it happens, the counter tube discharges and through a relay releases a hammer which shatters a small flask of hydrocyanic acid. If one has left this entire system to itself for an hour, one would say that the cat still lives if meanwhile no atom has decayed. The first atomic decay would have poisoned it. The [wavefunction] of the entire system would express this by having in it the living and the dead cat (pardon the expression) mixed or smeared out in equal parts.[10]

This scheme is illustrated here, in a slightly modified form.

Schrödinger said, in essence, that the quantum world of the Copenhagen interpretation differs radically from the world we experience on a daily basis. When we flip a coin and cover it with our

Erwin Schrödinger's experiment, with catnip. To avoid imagining a dead cat, I imagine a cat that is simultaneously high on catnip and not high on catnip, its state determined by the behavior of a radioactive atom. Drawing by Sarah Addy.

hand without looking, we know that it is already either heads or tails. According to Copenhagen, however, the coin would be in a wavelike state of heads and tails until we actually observe it. But this raises an additional problem: What causes a wavefunction to collapse, or, in other words, who does the observing? In a laboratory it is easy to imagine that wavefunction collapse is caused by a human scientist reading out the experimental results on an instrument, but this philosophically results in human beings playing a special role in the cosmos, an idea that science has moved inexorably away from for centuries.

Einstein approved of Schrödinger's criticism. In a 1950 letter he wrote:

> You are the only contemporary physicist, besides Laue, who sees that one cannot get around the assumption of reality—if only one is honest. Most of them simply do not see what sort of risky game they are playing with reality—reality as something independent of what is experimentally established. . . . This interpretation is, however, refuted, most elegantly by your system of radioactive atom + Geiger counter + amplifier + charge of gun powder + cat in a box, in which the [wavefunction] of the system contains the cat both alive and blown to bits. . . . Nobody really doubts that the presence or absence of the cat is something independent of the act of observation.[11]

Einstein preferred a gunshot for the cat's demise rather than a poison gas. We can only speculate about Schrödinger's choice of a cat to poison. He didn't have a cat himself, though he did have a beloved pet. During World War II he had a collie named Burshie (Laddie), who was a companion and source of comfort through many trials.

Despite the philosophical limitation, the Copenhagen interpretation of quantum physics is still taught in classrooms today and is still used as a practical model for what is going on in the quantum world. There are two simple reasons for this. First, it works well enough to interpret and explain all quantum laboratory experiments to date; the objections to it are primarily philosophical. Though this philosophy is very important, it does not slow the experimental study of the quantum world. Second, even after eighty years, nobody is entirely sure what to replace it with. One theory popular today, which involves an infinite set of parallel universes, was originally referred to as the many worlds theory of quantum mechanics. In this interpretation, Schrödinger's cat would be alive in one universe, dead in another, and the wave properties of quantum mechanics would represent an interaction of sorts between the universes. No way to test whether such parallel universes exist is yet known, but for many physicists the many worlds theory has become the preferred means of dealing with the strangeness of quantum physics.

Cats may have hinted at problems in our interpretation of the universe, but they also helped one astronomer broaden our understanding of it, at least with moral support. Before the twentieth century, it was generally thought that the Milky Way Galaxy, in which our solar system resides, was the entirety of the universe. Nebulae that could be observed with existing telescopes were thought to be clouds of gas that lurked within or immediately outside the galaxy. Then, in 1919, the American astronomer Edwin Hubble (1889–1953), having returned from World War I and one year as a student at Cambridge University, was offered a position on the staff at Mount Wilson Observatory, near Pasadena, California. Using the newly completed Hooker Telescope there, the world's largest telescope at the time, Hubble did extensive observations of the nebulae and convincingly demonstrated that they were much too far away to be considered a part of our galaxy; in fact, the indistinct blurs were galaxies themselves, incredibly distant from our own.

Hubble's discovery was announced to the world in the *New York Times* on November 23, 1924.[12] This publication marked the moment that the entire world learned that the universe is unfathomably larger than previously imagined. An excerpt of the article gives only a hint of that grandeur.

> Confirmation of the view that the spiral nebulae, which appear in the heavens as whirling clouds, are in reality distant stellar systems, or "island universes," has been obtained

> by Dr. Edwin Hubble of the Carnegie Institution's Mount Wilson Observatory, through investigations carried out with the observatory's powerful telescopes.
>
> The number of spiral nebulae, the observatory officials have reported to the institution, is very great, amounting to hundreds of thousands, and their apparent sizes range from small objects, almost star-like in character, to the great nebulae in Andromeda, which extends across an angle some 3 degrees in the heavens, about six times the diameter of the full moon.

The word *galaxy* is not used, but rather "island universes," for our galaxy was thought to be the entirety of the universe. Today astronomers estimate that there are some two trillion galaxies in the observable universe.

This monumental breakthrough was not Hubble's only major contribution to astronomy. In 1929, through careful observation, he noted that the speed at which distant galaxies are moving away from our own is proportional to their distance from us. This law, known as *Hubble's law*, is now a key tool for measuring distance in the cosmos.

Hubble worked at the Mount Wilson Observatory for the rest of his life. Astronomical observation can be a lonely occupation, with long hours spent late at night staring at the stars. Hubble and his wife, Grace, found a companion in 1946, a black furry kitten that Edwin immediately named Nicolas Copernicus after the Polish astronomer who in 1543 first correctly placed the Sun, rather than the Earth, in the center of our solar system.

Copernicus became a beloved member of the Hubble family. Edwin created a cat door for him: "All cats should have [one], it is necessary for their self-respect."[13] Pipe cleaners, Copernicus's

Edwin Hubble and Copernicus in March 1953. Image courtesy of the Observatories of the Carnegie Institution for Science Collection at the Huntington Library, San Marino, California.

preferred toy, were distributed around the house for him to play with.

Copernicus often "helped" Edwin with his work, as Grace wrote in her diary. "When E worked in the study at his big desk, Nicolas solemnly sprawled over as many pages as he could cover. 'He is helping me,' E explained. When he sat on E's lap, he purred differently, a slow, lion-like purr. . . . 'Is that your cat purring?' I would ask, and E would look up from his book, smile, and nod his head."[14] Edwin suffered a heart attack in 1949. In 1953, when he died of a blood clot in his brain, Copernicus was beside him in bed. He waited by the window for his master to return home for months afterward.

Copernicus may have been Hubble's research companion and good friend, but he didn't get quite as involved in scientific work as

Chester, the Siamese cat of Jack H. Hetherington of Michigan State University.

In 1975, Hetherington finished writing a paper that he intended to submit to the prestigious journal *Physical Review Letters* (*PRL*) as the sole author. Before submission, he passed the manuscript draft to a colleague to check one final time for any possible mistakes. Unfortunately, the colleague pointed out that Hetherington had written the paper referring to himself as "we" and "us," whereas *PRL* typically wanted single-author papers to be written with "I" and "me." In 1975, making that change would have involved retyping the entire manuscript on a typewriter, a bothersome task. So Hetherington added his cat as a co-author.

Hetherington's friends and colleagues would have recognized his cat Chester's name, so Hetherington expanded it to F. D. C. Willard. The "F. D." refers to the species name, *Felix domesticus,* while the "C" is, of course, "Chester." Willard was the name of Chester's father.

The paper was accepted and published in the November 24, 1975, issue of the journal. Hetherington didn't keep his co-author's identity secret for long. A day after the paper was published, he sent a message to his department chair about the ruse, and the chair, Truman Woodruff, sent back a note, offering to make Willard a visiting distinguished professor.

> Dear Jack:
> In response to your valued letter of 25 November: let me admit at once that if you had not written I should never have had the temerity to think of approaching so distinguished a physicist as F.D.C. Willard, F.R.S.C., with a view to interesting him in joining a university department like ours, which, after all, was not even rated among the best 30 in the 1969 Roose-Anderson

> study. Surely Willard can aspire to a connection with a more distinguished department.
>
> However, heartened by your view that he might conceivably deign to look with favor on the modest opportunity that we have to offer, I do beseech you—his friend and even collaborator—at the most propitious possible time (say, some evening when the brandy and cigars are going around) to raise the question with him (with all possible delicacy, I need hardly add). Can you imagine the universal jubilation if in fact Willard could be persuaded to join us, even if only as a Visiting Distinguished Professor?[15]

Word spread quickly among the rest of the physics community. As Hetherington himself wrote in a 1997 letter, "Shortly thereafter a visitor to MSU asked to talk to me, and since I was unavailable asked to talk with Willard. Everyone laughed and soon the cat was out of the bag."[16] Hetherington eventually distributed a few copies of the published paper to close friends and colleagues, signed by both authors. Apparently this signature, and the revelation that he was, in fact, a cat, kept Willard from being invited to at least one scientific conference. As Hetherington himself added, "It may or may not be significant that I did not receive an invitation to that conference either."[17]

Some people were clearly more delighted by the revelation than others. Jack Hetherington's wife, Marge, takes it as a point of personal pride that she can say that she slept with both authors of the paper—often simultaneously.

F. D. C. Willard went on to be the sole author of his own paper, "L'hélium 3 solide: un antiferromagnétique nucléaire," which was published in the French popular science magazine *La Recherche*

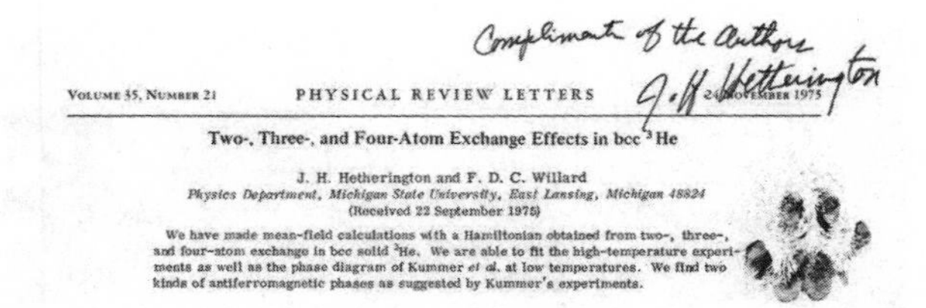

Compliment of the authors
J. H. Hetherington

VOLUME 35, NUMBER 21 PHYSICAL REVIEW LETTERS 24 NOVEMBER 1975

Two-, Three-, and Four-Atom Exchange Effects in bcc ^{3}He

J. H. Hetherington and F. D. C. Willard
Physics Department, Michigan State University, East Lansing, Michigan 48824
(Received 22 September 1975)

We have made mean-field calculations with a Hamiltonian obtained from two-, three-, and four-atom exchange in bcc solid ^{3}He. We are able to fit the high-temperature experiments as well as the phase diagram of Kummer *et al.* at low temperatures. We find two kinds of antiferromagnetic phases as suggested by Kummer's experiments.

"Compliments of the authors": a copy of the published paper signed by both authors. Courtesy of Jack H. Hetherington.

in September 1980. Evidently the human co-authors of the paper could not come to an agreement about some of the details of their manuscript, so they pushed all the blame and responsibility onto their feline colleague.

Hetherington retired from Michigan State University after thirty-five years but remains active. He now divides his time between Michigan and France and not only does work for an engineering company but also explores the artistic possibilities in the representation of mathematical functions.[18]

The American Physical Society, which publishes *Physical Review Letters*, had a sense of humor about the whole affair. On April 1, 2014, it announced on its website a new "open access initiative" targeting future feline researchers.

> APS is proud to announce a new open access initiative designed to further extend the benefits of open access to a broader set of authors. The new policy, effective today, makes all papers authored by cats freely available. This open-minded update is a natural extension of APS's leadership in both open access and pet publishing. As early as 1975, APS began publishing papers with feline authors, most notably the

The actual F. D. C. Willard. Courtesy of Jack H. Hetherington.

> contribution by one F. D. C. Willard [J. H. Hetherington and F. D. C. Willard, Two-, Three-, and Four-Atom Exchange Effects in bcc 3He, Phys. Rev. Lett. 35, 1442 (1975)]. Going forward, only single author papers will be considered. APS hopes to evaluate allowing publication by canine authors in the near future. Not since Schrödinger has there been an opportunity like this for cats in physics.

F. D. C. Willard was able to publish in one of the most prestigious physics journals in the world but was never able to earn an academic degree. Other cats, however, have been successful in this regard, providing an important service to academia.

Around the year 2001, one Zoe D. Katze was accredited to perform hypnotherapy, becoming Dr. Katze. Zoe was the feline companion of psychologist Steven Eichel, who earned the certification for Katze remotely through the American Psychotherapy Association. The purpose of the exercise was to show, in the most dramatic and absurd manner possible, how easily anyone could become licensed to work with patients. Another accredited cat is Henrietta, the feline companion of science journalist Ben Goldacre. Henrietta earned a diploma in nutrition in 2004 from the American Association of Nutritional Consultants, an especially impressive feat since she had died a year earlier. Such investigations—and there are many others—have been used to highlight shady "diploma mills" that will award a degree to anyone for the right price.

Will a cat ever earn a degree in physics? If recent research is any indication, they know more about the subject than we give them credit for. In 2016 researchers at Kyoto University undertook a study of how well cats understand causality: the relationship between a cause and an effect.[19] To do so, they built a can that contained both an electromagnet and three metal balls that could rattle around inside. With the magnet active, the balls would be fixed in place and would not rattle when the can was shaken, and would not fall out of the can when it was turned over. With the magnet off, the balls would rattle noisily and could fall out.

The goal of the experiment was to test whether cats would connect the rattling sound with the expectation that the balls would fall out of the can. The magnet allowed four scenarios to play out. With the magnet off the researchers could test "congruent conditions"—namely, a filled can rattles and balls fall out, or an empty can doesn't rattle and no balls fall out. With the magnet active half the time, they could test "incongruent conditions"—namely, a filled can rattles but balls do not fall out, or a filled can doesn't rattle, but balls do fall out.

The researchers filmed thirty domestic cats—eight house cats and twenty-two cats from cat cafés—to gauge their responses to congruent and incongruent conditions. The cats were allowed to investigate the can after each event if they so chose. It was found that cats looked at the container for longer periods of time when the demonstration

A member of my extended kitty family, Sophie, contemplates string theory.

was incongruent with normal expectations. That is, if the can rattled but no balls fell out, or the can didn't rattle but balls fell out, the cats were more curious as to what was going on. According to the researchers, their "results suggest that cats used a causal-logical understanding of auditory stimuli to predict the appearance of invisible objects." On reflection, this is not surprising, for cats who are out on the hunt will be more successful if they are able to connect a set of sounds with the presence of a hidden prey animal.

Some news outlets had a more sensational take on the results. An article about the work at the Smithsonian led with the headline "Cats Are Adorable Physicists."[20] In my own experience, as the accompanying photograph demonstrates, some cats are already leaning into the subject.

Notes

Chapter 1. Famous Physicists' Fascination with Falling Felines

1. Campbell and Garnett, *The Life of James Clerk Maxwell,* p. 499. Maxwell probably meant "feet," not "inches," since two feet is the distance determined in research as the minimum height at which a cat can turn over, such as that done by Fiorella Gambale and published in the semi-satirical *Annals of Improbable Research* (Gambale, "Does a Cat Always Land on Its Feet?"). Since Maxwell's statement is in a nontechnical handwritten letter to his wife, he either didn't notice or didn't feel the need to correct the typo. His wife was, in fact, a skilled scientist and helped her husband with some of his experimental studies.

2. Stokes, *Memoir and Scientific Correspondence,* p. 32.

3. G. R. Tomson, excerpt from "To My Cat," in Tomson, *Concerning Cats.*

Chapter 2. The (Solved?) Puzzle of the Falling Cat

1. Ross, *The Book of Cats.*

2. The biographical data here come largely from Woods, "Stables, William Gordon."

3. Stables, *"Cats,"* p. 391.

4. Stables, *From Ploughshare to Pulpit,* pp. 126–127.

5. Battelle, *Premières Leçons d'Histoire Naturelle,* p. 48. My translation.

6. Defieu, *Manuel Physique,* pp. 69–70. My translation.

7. Quitard, *Dictionnaire Étymologique, Historique et Anecdotique des Proverbes,* p. 211. My translation.

8. Errard, *La Fortification Démonstrée et Réduicte en Art.*

9. Hutton, *A Mathematical and Philosophical Dictionary,* vol. 2, pp. 200–201; "Éloge de M. Parent."

10. Parent, "Sur les corps qui nagent dans des liqueurs."

11. I have personal knowledge of this as an experienced skydiver.

12. As described in Grayling, *Descartes,* p. 160. It is unclear how old this legend is, and much less clear whether it is true.

13. Bossewell and Legh, *Workes of Armorie,* p. F 0.56.

14. Garnett, *The Women of Turkey and Their Folk-Lore,* pp. 516–517.

15. Chittock, *Cats of Cairo.*

16. Stables, "*Cats,*" pp. 3–6.

Chapter 3. Horses in Motion

1. Lankester, "The Problem of the Galloping Horse."

2. Renner, *Pinhole Photography,* p. 4.

3. Before "science" or "natural philosophy" were recognized disciplines, unusual tricks involving natural phenomena were considered "natural magic."

4. *Dictionnaire Technologique,* p. 391. My translation.

5. Potonniée, *The History of the Discovery of Photography,* pp. 47–48.

6. Potonniée, *The History of the Discovery of Photography,* pp. 97–99.

7. Potonniée, *The History of the Discovery of Photography,* pp. 114–115 (quote, p. 114).

8. Potonniée, *The History of the Discovery of Photography,* pp. 160–161.

9. Talbot's wife, Constance Fox Talbot, was more supportive of her husband's experiments than Daguerre's wife was of his. Constance is credited as the first woman to take a photograph, for she briefly experimented with photography around 1839. Talbot also taught photography to Anna Atkins (1799–1871), who went on to publish the first book including photographic images, *Photographs of British Algae: Cyanotype Impressions,* in 1843.

10. Gihon, "Instantaneous Photography."

11. A popular history of Muybridge and his work is Rebecca Solnit's *River of Shadows.*

12. Helios, "A New Sky Shade."

13. Solnit, *River of Shadows,* p. 80; emphasis mine.

14. "A Horse's Motion Scientifically Determined."

15. Lankester, "The Problem of the Galloping Horse."

16. Personal and Political, *Philadelphia Inquirer,* August 6, 1881.

17. Marey, "Sur les allures du cheval reproduites," p. 74.

Chapter 4. Cats on Film

1. Much of the information on Marey comes from Marta Braun's definitive biography, *Picturing Time.* For the quotation see p. 2.

2. Toulouse, "Nécrologie–Marey." My translation.

3. Nadar, "Le nouveau president." Translation by Jana Sloan van Geest.

4. Marey, *Animal Mechanism,* p. 8.

5. Muybridge, "Photographies instantanées des animaux en mouvement." My translation.

6. This is a translation of an account from *Le Globe,* Paris, taken from the *Daily Alta California* of November 16, 1881.

7. This anecdote is widely reported, but I have been unable to track down a solid source for it despite its plausibility.

8. Marey, "Des mouvements que certains animaux."

9. "Why Cats Always Land on Their Feet."

Chapter 5. Going Round and Round

1. A detailed history can be found in Coopersmith, *Energy, the Subtle Concept.*

2. "Perpetual Motion."

3. Newton, *The Mathematical Principles of Natural Philosophy.*

4. Rankine, *Manual of Applied Mechanics.*

5. "Par-ci, par-là," p. 706. Translation by Jana Sloan van Geest.

6. Anderson, "Analyzing Motion," p. 490.

7. Delaunay, *Traité de Méchanique Rationnelle,* p. 450. My translation.

8. Tait, "Clerk-Maxwell's Scientific Work."

9. *The Nation,* November 29, 1894, pp. 409–410.

10. Guyou, "Note relative à la communication de M. Marey."

11. Lévy, "Observations sur le principe des aires"; Deprez, "Sur un appareil. servant à mettre en évidence certaines conséquences du théorème des aires."

12. "Photographs of a Tumbling Cat," *Nature,* 1849, pp. 80–81.

13. Routh, *Dynamics of a System of Rigid Bodies,* p. 237.

14. Lecornu, "Sur une application du principe des aires."

15. W. Wright, *Flying.*

16. "Meissonier and Muybridge," *Sacramento Daily Union,* December 28, 1881.

Chapter 6. Cats Rock the World

1. Peano, "Il principio delle aree e la storia d'un gatto."

2. Fredrickson, "The Tail-Less Cat in Free-Fall."

3. Many pieces of information about Peano used in this chapter are from H. C. Kennedy's *Peano: Life and Works of Giuseppe Peano.*

4. Peano, "Sur une courbe."

5. A discussion of Chandler's discoveries and his influence is found in Carter and Carter, "Seth Carlo Chandler Jr."

6. Chandler, "On the Variation of Latitude, I"; Chandler, "On the Variation of Latitude, II."

7. "Notes on Some Points Connected with the Progress of Astronomy during the Past Year." For Simon Newcomb's response at the meeting, see Newcomb, "On the Dynamics of the Earth's Rotation."

8. Peano, "Sopra la spostamento del polo sulla terra." My translation.

9. Volterra, "Sulla teoria dei moti del polo terrestre."

10. Biographical information is taken from E. T. Whittaker's obituary of Volterra; see Whittaker, "Vito Volterra."

11. Volterra, "Sulla teoria dei movimenti del polo terrestre."

12. "Adunanza del 5 maggio 1895"; Volterra, "Sui moti periodici del polo terrestre."

13. Volterra, "Sulla teoria dei moti del polo nella ipotesi della plasticità terrestre"; Volterra, "Osservazioni sulla mia Nota."

14. Peano, "Sul moto del polo terrestre" (1895).

15. Volterra, "Sulla rotazione di un corpo in cui esistono sistemi ciclici."

16. Volterra, "Sul moto di un sistema nel quale sussistono moti interni variabili." Translation by H. C. Kennedy.

17. Peano, "Sul moto di un sistema nel quale sussistono moti interni variabili." My translation. The two "Notes" referred to here by Peano are "Sopra la spostamento del polo sulla terra" and "Sul moto del polo terrestre." When presented to the Academy, they had the same title, but when they were published in print, they were given new and different titles.

18. Volterra, "Il Presidente Brioschi dà comunicazione della seguente lettera, ricevuta dal Corrispondente V. Volterra." Translation by H. C. Kennedy.

19. Peano, "Sul moto del polo terrestre" (1896).

20. Malkin and Miller, "Chandler Wobble"; Gross, "The Excitation of the Chandler Wobble."

Chapter 7. The Cat-Righting Reflex

1. Galileo, *Dialogue Concerning the Two Chief World Systems,* pp. 186–187.

2. A. Einstein, "Excerpt from Essay by Einstein on Happiest Thought in His Life," *New York Times,* March 28, 1972.

3. I can confirm this personally. The moment after jumping out of a hot air balloon, before air resistance becomes significant, one is truly in freefall and weightless.

4. Hall, "Medulla Oblongata and Medulla Spinalis."

5. Bell, "Nerves of the Orbit."

6. See, for example, Gary Busey's character "Mr. Joshua" in the first *Lethal Weapon* movie.

7. Much of Sherrington's biographical information offered here comes from his obituary. Liddell, "Charles Scott Sherrington."

8. The story of Sherrington's parentage is still a matter of debate, though the description provided here is apparently the most accepted.

9. Sherrington, "Note on the Knee-Jerk."

10. Sherrington, "On Reciprocal Innervation of Antagonistic Muscles."

11. Sherrington, *Inhibition as a Coordinative Factor.*

12. Levine, "Sherrington's 'The Integrative Action of the Nervous System.' "

13. Weed, "Observations upon Decerebrate Rigidity."

14. Muller and Weed, "Notes on the Falling Reflex of Cats."

15. Sherrington, *The Integrative Action of the Nervous System,* p. 302.

16. Biographical information about Magnus comes from the biography written by his son, Otto Magnus: *Rudolf Magnus—Physiologist and Pharmacologist.*

17. R. Magnus, "Animal Posture."

18. R. Magnus, "Wie sich die fallende Katze in der Luft umdreht." My translation.

19. Rademaker and ter Braak, "Das Umdrehen der fallenden Katze in der Luft."

20. The limited information available on Rademaker's history comes from his 1957 obituary by H. Verbiest, "In Memoriam."

21. Brindley is also infamous for a presentation at a urology conference in Las Vegas in 1983, which shocked listeners but is nevertheless considered an important milestone in erectile dysfunction treatment.

22. Brindley, "How Does an Animal Know the Angle?"; Brindley, "Ideal and Real Experiments to Test the Memory Hypothesis."

23. Kan et al., "Biographical Sketch, Giles Brindley, FRS."

Chapter 8. Cats . . . in . . . Space!

1. The full video, *Biometrics Research,* can be seen online at the Aeronautical Systems Division Motion Picture Section, https://www.youtube.com/watch?v=HwRdc v8azvk. It is often mistakenly said to have been filmed in 1947, because of the date of the archive series. The cats appear in the "weightless" section, starting at 3:00.

2. An excellent discussion of the early history of spaceflight can be found in Amy Shira Teitel's *Breaking the Chains of Gravity.*

3. For example, in March 2016, American astronaut Scott Kelly ended a full year in space aboard the International Space Station. His twin brother had stayed on Earth, which provided a unique opportunity to see what sort of biological changes can happen to an individual in space.

4. Haber, "The Human Body in Space."

5. Gauer and Haber, *Man under Gravity-Free Conditions,* pp. 641–644.

6. Haber and Haber, "Possible Methods of Producing the Gravity-Free State."

7. In 2016, the band OK-Go shot a music video for their song "Upside Down & Inside Out" inside a plane performing this weightless up-and-down maneuver. In the video, you can see the moments where the band pauses their weightless antics when the pilot comes out of the dive.

8. Gerathewohl, "Subjects in the Gravity-Free State."

9. Gerathewohl, "Subjects in the Gravity-Free State."

10. Ballinger, "Human Experiments in Subgravity and Prolonged Acceleration."

11. Henry et al., "Animal Studies of the Subgravity State."

12. Gazenko et al., "Harald von Beckh's Contribution."

13. Von Beckh, "Experiments with Animals and Human Subjects."

14. Gerathewohl and Stallings, "The Labyrinthine Postural Reflex."

15. Schock, "A Study of Animal Reflexes."

16. Experiments in these planes were described in two publications by E. L. Brown, "Human Performance and Behavior during Zero Gravity" and "Research on Human Performance during Zero Gravity."

17. E. L. Brown, "Research on Human Performance during Zero Gravity."

18. "Pioneer Space Group to Mark Lab Foundation," Associated Press, February 8, 1959.

19. Kulwicki, Schlei, and Vergamini, *Weightless Man.*

20. Whitsett, *Some Dynamic Response Characteristics of Weightless Man.* It is tempting to view this as a stereotypical example of scientists and engineers being overly

analytical and nonpoetic. Knowing many such people, however, I strongly suspect the authors were laughing a lot when they wrote this sentence.

21. Stepantsov, Yeremin, and Alekperov, *Maneuvering in Free Space.*

22. Robe and Kane, "Dynamics of an Elastic Satellite—I."

23. Smith and Kane, "On the Dynamics of the Human Body in Free Fall."

24. Kane and Scher, "A Dynamical Explanation of the Falling Cat Phenomenon."

25. "A Copycat Astronaut," *Life Magazine,* June 30, 1968.

26. Kane and Scher, "Human Self-Rotation by Means of Limb Movements."

Chapter 9. Cats as Keepers of Mysteries

1. *Kentish Times,* March 4, 1825.

2. Bleecker, "Jungfrau Spaiger's Apostrophe to Her Cat."

3. Thompson, "Spiders and the Electric Light."

4. Robinson, "The High Rise Trauma Syndrome in Cats."

5. Whitney and Mehlhaff, "High-Rise Syndrome in Cats."

6. A. Parachini, "They Land on Little Cat Feet," *Los Angeles Times,* December 28, 1987; "On Landing Like a Cat: It Is a Fact," *New York Times,* August 22, 1989.

7. Papazoglou et al., "High-Rise Syndrome in Cats"; Merbl et al., "Findings in Feline High Rise Syndrome in Israel."

8. Vnuk et al., "Feline High-Rise Syndrome."

9. Skarda, "Cat Survives 19-Story Fall by Gliding Like a Flying Squirrel."

10. Binette, "Cat Is Unharmed after 26 Story Fall from High Rise Building."

11. There are arguments that the plane may have been much lower when the explosion happened, maybe 10,000 feet, but this is irrelevant to the story of survival, since terminal velocity for humans is achieved after falling 1,500 feet.

12. Studnicka, Slegr, and Stegner, "Free Fall of a Cat—Freshman Physics Exercise."

13. A. Parachini, "They Land on Little Cat Feet," *Los Angeles Times,* December 28, 1987.

14. Brehm, "The Surprising Physics of Cats' Drinking."

15. Stratton, "Harold Eugene Edgerton."

16. Vandiver and Kennedy, "Harold Eugene Edgerton."

17. Thone, "Right Side Up." Comma added.

18. K. Bruillard, "A Cat's Sandpapery Tongue Is Actually a Magical Detangling Hairbrush," *Washington Post,* November 29, 2015, online.

19. Gaal, "Cat Tongues Are the Ultimate Detanglers."

Chapter 10. Rise of the Robotic Cats

1. "Dante Spends Another Night inside Volcano," *Ukiah Daily Journal,* August 9, 1994, p. 17; "Dante II Bound for Museum," *Daily Sitka Sentinel,* October 20, 1994, p. 3.

2. "Robot to Be Turned to Inspection of Volcano," *Daily Sitka Sentinel,* July 6, 1994, p. 7.

3. For further discussion of biorobotics see Beer, "Biologically Inspired Robotics."

4. Deprez, "Sur un appareil servant à mettre en évidence certaines conséquences du théorèeme des aires."

5. "The Steam Man."

6. "Hercules, the Iron Man," *Washington Standard,* November 22, 1901, p. 1. Period deleted.

7. Walter, "An Imitation of Life."

8. Holland, "The First Biologically Inspired Robots."

9. Vincent et al., "Biomimetics."

10. Ballard et al., "George Charles Devol, Jr."

11. Brooks, "New Approaches to Robotics."

12. Triantafyllou and Triantafyllou, "An Efficient Swimming Machine."

13. Beer et al., "Biologically Inspired Approaches to Robotics."

14. Espenschied et al., "Biologically Based Reflexes in a Hexapod Robot."

15. Espenschied et al., "Leg Coordination Mechanisms in the Stick Insect."

16. Kim et al., "Whole Body Adhesion."

17. Galli, "Angular Momentum Conservation and the Cat Twist"; Frohlich, "The Physics of Somersaulting and Twisting."

18. Arabyan and Tsai, "A Distributed Control Model for the Air-Righting Reflex."

19. O'Leary and Ravasio, "Simulation of Vestibular Semicircular Canal Responses."

20. Arabyan and Tsai, "A Distributed Control Model for the Air-Righting Reflex."

21. Ge and Chen, "Optimal Control of a Nonholonomic Motion"; Putterman and Raz, "The Square Cat"; Kaufman, "The Electric Cat"; Zhen et al., "Why Can a Free-Falling Cat Always Manage to Land Safely?"

22. Davis et al., "A Review of Self-Righting Techniques for Terrestrial Animals."

23. In many martial arts, students are taught to connect with the ground with an arm first when rolling or falling; it can be used to guide the rest of the body safely to landing.

24. Davis et al., "A Review of Self-Righting Techniques for Terrestrial Animals."

25. Jusufi et al., "Aerial Righting Reflexes in Flightless Animals."

26. Jusufi et al., "Active Tails Enhance Arboreal Acrobatics in Geckos"; Jusufi et al., "Righting and Turning in Midair Using Appendage Inertia."

27. Libby et al., "Tail-Assisted Pitch Control in Lizards, Robots and Dinosaurs."

28. Walker, Vierck, and Ritz, "Balance in the Cat."

29. Shield, Fisher, and Patel, "A Spider-Inspired Dragline."

30. Dunbar, "Aerial Maneuvers of Leaping Lemurs."

31. Bergou et al., "Falling with Style."

32. Yamafuji, Kobayashi, and Kawamura, "Elucidation of Twisting Motion of a Falling Cat"; Kawamura, "Falling Cat Phenomenon and Realization by Robot."

33. Shields et al., "Falling Cat Robot Lands on Its Feet."

34. Bingham et al., "Orienting in Mid-Air through Configuration Changes"; Wagstaff, "Purr-plexed?"

35. Sadati and Meghdari, "Singularity-Free Planning for a Robot Cat Freefall."

36. Pope and Niemeyer, "Falling with Style."

37. Zhao, Li, and Feng, "Effect of Swing Legs on Turning Motion."

38. Haridy, "Boston Dynamics' Atlas Robot."

39. H. Pettit. "Scientists Create an AI Robot CAT That Helps Keep the Elderly Company and Reminds Them to Take Their Medication," *Daily Mail,* December 19, 2017, online.

Chapter 11. The Challenges of Cat-Turning

1. L. E. Brown, "Seeing the Elephant."

2. Kawamura, "Falling Cat Phenomenon and Realization by Robot."

3. Franklin, "How a Falling Cat Turns Over in the Air"; Benton, "How a Falling Cat Turns Over."

4. McDonald, "The Righting Movements of the Freely Falling Cat"; McDonald, "How Does a Falling Cat Turn Over?" (quotation).

5. Mpemba and Osborne, "Cool?"

6. Mpemba and Osborne, "Cool?"

7. Aristotle, *Meteorology.*

8. Bacon, *Novum Organum,* p. 319; Descartes, *Discourse on Method,* p. 268.

9. A good alternative discussion of the Mpemba effect and its historical development is in Ouellette, "When Cold Warms Faster Than Hot."

10. Wojciechowski, Owczarek, and Bednarz, "Freezing of Aqueous Solutions Containing Gases."

11. Auerbach, "Supercooling and the Mpemba Effect"; Brownridge, "When Does Hot Water Freeze Faster Then Cold Water?"

12. Katz, "When Hot Water Freezes before Cold."

13. Burridge and Linden, "Questioning the Mpemba Effect"; Lu and Raz, "Nonequilibrium Thermodynamics of the Markovian Mpemba Effect"; Lasanta et al., "When the Hotter Cools More Quickly."

14. "How Do Cats Always Land on Their Feet?"; "Leopard Cub Falling Out of a Tree."

15. McDonald, "How Does a Cat Fall on Its Feet?"; McDonald, "How Does a Man Twist in the Air?"

16. Biesterfeldt, "Twisting Mechanics II"; Frohlich, "Do Springboard Divers Violate Angular Momentum Conservation?"; Yeadon, "The Biomechanics of Twisting Somersaults"; Dapena, "Contributions of Angular Momentum and Catting."

17. Frohlich, "Do Springboard Divers Violate Angular Momentum Conservation?"

Chapter 12. Falling Felines and Fundamental Physics

1. There is another solution to the "bear puzzle"; in fact, there are an infinite number of solutions. The puzzle is posed and explained by classic puzzler Martin Gardner in *My Best Mathematical and Logic Puzzles.*

2. Foucault, "Physical Demonstration of the Rotation of the Earth."

3. "Foucault, the Academician."

4. You might argue, "What if I walk forward ten feet, then sidestep left ten feet, then walk backward ten feet, then sidestep right ten feet? The pendulum will not change direction." You are correct, but in our example—and in physics—we restrict ourselves to what is known as *parallel transport,* which effectively means we always keep our cafeteria tray oriented the same way with respect to the direction we're walking.

5. Berry and Wilkinson, "Diabolical Points in the Spectra of Triangles."

6. Berry, "Geometric Phase Memories."

7. Berry, "Quantal Phase Factors Accompanying Adiabatic Changes."

8. Mead and Truhlar, "On the Determination of Born-Oppenheimer Nuclear Motion Wave Functions."

9. Pancharatnam, "Generalized Theory of Interference."

10. We cannot directly see the oscillation of visible light, which wiggles at roughly a million billion times per second.

11. Berry, "The Adiabatic Phase and Pancharatnam's Phase."

12. Marsden, Montgomery, and Ratiu, "Reduction, Symmetry, and Phase in Mechanics."

13. Batterman, "Falling Cats, Parallel Parking, and Polarized Light."

14. The complete path around a circle is 360 degrees, which is equal to 2π radians.

15. Cats that are long and skinny are represented by a geometric shape more like an egg.

16. Montgomery, "Gauge Theory of the Falling Cat"; Iwai, "Classical and Quantum Mechanics of Jointed Rigid Bodies."

17. Chryssomalakos, Hernández-Coronado, and Serrano-Ensástiga, "Do Free-Falling Quantum Cats Land on Their Feet?"

Chapter 13. Scientists and Their Cats

1. Levenson, *Newton and the Counterfeiter*, p. 8.

2. "Philosophy and Common Sense."

3. J. M. Wright, *Alma Mater,* pp. 15–18. Dash added and typo fixed.

4. Graves, *Life of Sir William Rowan Hamilton,* vol. 3, pp. 235–236.

5. Diecke, "Robert Williams Wood."

6. Seabrook, *Doctor Wood.*

7. "A Story of Youth Told by Age: Dedicated to Miss Pola Fotich, by Its Author Nikola Tesla," in *Tesla: Master of Lightning.*

8. This often-quoted statement comes from a letter from Albert Einstein to Michele Besso, December 12, 1951. My thanks to Dr. Jens Foell for helping with a translation.

9. Letter to Ilse Kayser-Einsetin and Rudolf Kayser, May 21, 1924, in *The Collected Papers of Albert Einstein,* vol. 14, p. 214.

10. Schrödinger, *The Present Situation in Quantum Mechanics,* pp. 152–167.

11. Maxwell, "Induction and Scientific Realism," p. 290.

12. "Finds Spiral Nebulae Are Stellar Systems," *New York Times,* November 23, 1924.

13. Wehrey, "Hubble and Copernicus."

14. Wehrey, "Hubble and Copernicus."

15. Woodruff, "WoodruffLetter."

16. Hetherington, "Letter to Ms. Lubkin."

17. Weber, *More Random Walks in Science.*

18. C. Opper, "Jack Hetherington Finds Beauty in Data," *Lansing City Pulse,* June 8, 2016.

19. Takagi et al., "There's No Ball without Noise."

20. Blakemore, "Cats Are Adorable Physicists."

Bibliography

"Adunanza del 5 maggio 1895." *Atti della Reale Accademia delle scienze di Torino,* 30:513–514, 1895.

Anderson, A. "Analyzing Motion." *Pearson's Magazine,* 13:484–491, 1902.

Arabyan, A., and Derliang Tsai. "A Distributed Control Model for the Air-Righting Reflex of a Cat." *Biological Cybernetics,* 79:393–401, 1998.

Aristotle. *Meteorology.* Princeton University Press, Princeton, NJ, 1984.

Auerbach, D. "Supercooling and the Mpemba Effect: When Hot Water Freezes Quicker Than Cold." *American Journal of Physics,* 63:882–885, 1995.

Bacon, F. *Novum Organum.* William Pickering, London, 1844.

Ballard, L. A., S. Šabanović, J. Kaur, and S. Milojević. "George Charles Devol, Jr." *IEEE Robotics and Automation Magazine,* pages 114–119, December 2012.

Ballinger, E. R. "Human Experiments in Subgravity and Prolonged Acceleration." *Journal of Aviation Medicine,* 23:319–321, 1952.

Battelle, G. M. *Premières Leçons d'Histoire Naturelle: Animaux Domestiques.* Hachette, Paris, 1836.

Batterman, R. W. "Falling Cats, Parallel Parking, and Polarized Light." *Studies in History and Philosophy of Modern Physics,* 34:527–557, 2003.

Beer, R. D. "Biologically Inspired Robotics." *Scholarpedia,* 4(4):1531, 2009. Revision #91061.

Beer, R. D., R. D. Quinn, H. J. Chiel, and R. E. Ritzmann. "Biologically Inspired Approaches to Robotics." *Communications of the ACM,* 40:31–38, 1997.

Bell, C. "Second Part of the Paper on the Nerves of the Orbit." *Philosophical Transactions of the Royal Society of London,* 113: 289–307, 1823.

Benton, J. R. "How a Falling Cat Turns Over." *Science,* 35:104–105, 1912.

Bergou, A. J., S. M. Swartz, H. Vejdani, D. K. Riskin, L. Reimnitz, G. Taubin, and K. S. Breuer. "Falling with Style: Bats Perform Complex Aerial Rotations by Adjusting Wing Inertia." *PLOS Biology,* 13:e1002297, 2015.

Berry, M. V. "The Adiabatic Phase and Pancharatnam's Phase for Polarized Light." *Journal of Modern Optics,* 34:1401–1407, 1987.

Berry, M. V. "Geometric Phase Memories." *Nature Physics,* 6:148–150, 2010.

Berry, M. V. "Quantal Phase Factors Accompanying Adiabatic Changes." *Proceedings of the Royal Society of London A,* 392:45–57, 1984.

Berry, M. V., and M. Wilkinson. "Diabolical Points in the Spectra of Triangles." *Proceedings of the Royal Society of London A,* 392:15–43, 1984.

Biesterfeldt, H. J. "Twisting Mechanics II." *Gymnastics,* 16:46–47, 1974.

Binette, K. H. "Cat Is Unharmed after 26 Story Fall from High Rise Building." *Life with Cats,* February 17, 2015. https://www.lifewithcats.tv/2015/02/17/cat-is-unharmed-after-26-story-fall-from-high-rise-building/.

Bingham, J. T., J. Lee, R. N. Haksar, J. Ueda, and C. K. Liu. "Orienting in Mid-Air through Configuration Changes to Achieve a Rolling Landing for Reducing Impact after a Fall." In *IEEE/RSJ International Conference on Intelligent Robots and Systems,* pages 3610–3617, 2014.

Blakemore, E. "Cats Are Adorable Physicists." *Smithsonian,* June 16, 2016. Online.

Bleecker, A. "Jungfrau Spaiger's Apostrophe to Her Cat." In S. Kettell, ed., *Specimens of American Poetry.* S. G. Goodrich, Boston, 1829.

Bossewell, John, and Gerard Legh. *Workes of Armorie: Deuyded into three bookes, entituled, the Concordes of armorie, the Armorie of honor, and of Coates and creastes.* In aedibus Richardi Totelli, London, 1572.

Braun, M. *Picturing Time.* University of Chicago Press, Chicago, 1992.

Brehm, D. "The Surprising Physics of Cats' Drinking." *MIT News,* November 12, 2010. Online.

Brindley, G. S. "How Does an Animal That Is Dropped in a Non-Upright Posture Know the Angle through Which It Must Turn in the Air So That Its Feet Point to the Ground?" *Journal of Physiology,* 180:20–21P, 1965.

Brindley, G. S. "Ideal and Real Experiments to Test the Memory Hypothesis of Righting in Free Fall." *Journal of Physiology,* 184:72–73P, 1966.

Brindley, G. S. "The Logical Bassoon." *The Galpin Society Journal,* 21:152–161, 1968.

Brooks, R. A. "New Approaches to Robotics." *Science,* 253:1227–1232, 1991.

Brown, E. L. "Human Performance and Behavior during Zero Gravity." In E. T. Benedikt, ed., *Weightlessness—Physical Phenomena and Biological Effects.* Springer, New York, 1961.

Brown, E. L. "Research on Human Performance during Zero Gravity." In G. Finch, ed., *Air Force Human Engineering, Personnel, and Training Research.* National Academy of Sciences, Washington, DC, 1960.

Brown, Rev. L. E. "Seeing the Elephant." *Bulletin of Comparative Medicine and Surgery,* 2:1–4, 1916.

Brownridge, J. D. "When Does Hot Water Freeze Faster Than Cold Water? A Search for the Mpemba Effect." *American Journal of Physics,* 79:78–84, 2011.

Burridge, H. C., and P. F. Linden. "Questioning the Mpemba Effect: Hot Water Does Not Cool More Quickly Than Cold." *Scientific Reports,* 6:37665, 2016.

Campbell, L., and W. Garnett. *The Life of James Clerk Maxwell,* page 499. Macmillan, London, 1882.

Carter, M. S., and W. E. Carter. "Seth Carlo Chandler Jr.: The Discovery of Variation of Latitude." In *Polar Motion: Historical and Scientific Problems,* volume 208 of *ASP Conference Series,* pages 109–122, 2000.

Chandler, S. C. "On the Variation of Latitude, I." *Astronomical Journal,* 248:59–61, 1891.

Chandler, S. C. "On the Variation of Latitude, II." *Astronomical Journal,* 249:65–70, 1891.

Chittock, L. *Cats of Cairo: Egypt's Enduring Legacy.* Abbeville, New York, 2001.

Chryssomalakos, C., H. Hernández-Coronado, and E. Serrano-Ensástiga. "Do Free-Falling Quantum Cats Land on Their Feet?" *Journal of Physics A,* 48:295301, 2015.

Coopersmith, J. *Energy, the Subtle Concept.* Oxford University Press, Oxford, revised edition, 2015.

"A Copycat Astronaut." *Life Magazine,* June 30, 1968.

Dapena, J. "Contributions of Angular Momentum and Catting to the Twist Rotation in High Jumping." *Journal of Applied Biomechanics,* 13:239–253, 1997.

Davis, M., C. Gouinand, J-C. Fauroux, and P. Vaslin. "A Review of Self-Righting Techniques for Terrestrial Animals." In *International Workshop for Bio-inspired Robots,* 2011. Online.

Defieu, J. F. *Manuel Physique.* Regnault, Lyon, 1758.

Delaunay, M. C. *Traité de Méchanique Rationnelle.* Langlois and Leclercq, Paris, 1856.

Deprez, M. "Sur un appareil servant à mettre en évidence certaines conséquences du théorème des aires." *Comptes Rendus,* 119:767–769, 1894.

Descartes, R. *Discourse on Method, Optics, Geometry, and Meteorology.* Translated by P. J. Olscamp. Bobbs-Merrill, Indianapolis, 1965.

Dictionnaire Technologique. Chez Thomine et Fortic, Paris, 1823.

Diecke, G. H. "Robert Williams Wood, 1868–1955." *Biographical Memoirs of Fellows of the Royal Society,* 2:326–345, 1956.

Dunbar, D. C. "Aerial Maneuvers of Leaping Lemurs: The Physics of Whole-Body Rotations while Airborne." *American Journal of Primatology,* 16:291–303, 1988.

Einstein, A. *The Collected Papers of Albert Einstein,* volume 14 (English Translation Supplement). Princeton University Press, Princeton, NJ, 2015.

"Éloge de M. Parent." *Histoire de l'Academie Royale,* pp. 88–93. 1716.

Errard, J. *La Fortification Démonstrée et Réduicte en Art.* Paris, 1600.

Espenschied, K. S., R. D. Quinn, H. J. Chiel, and R. D. Beer. "Leg Coordination Mechanisms in the Stick Insect Applied to Hexapod Robot Locomotion." *Adaptive Behavior,* 1:455–468, 1993.

Espenschied, K. S., R. D. Quinn, R. D. Beer, and H. J. Chiel. "Biologically Based Distributed Control and Local Reflexes Improve Rough Terrain Locomotion in a Hexapod Robot." *Robotics and Autonomous Systems,* 18:59–64, 1996.

Foucault, L. "Physical Demonstration of the Rotation of the Earth by Means of the Pendulum." *Journal of the Franklin Institute,* 21: 350–353, 1851.

"Foucault, the Academician." *Putnam's Monthly,* 8:416–421, 1857.

Franklin, W. S. "How a Falling Cat Turns Over in the Air." *Science,* 34:844, 1911.

Fredrickson, J. E. "The Tail-Less Cat in Free-Fall." *Physics Teacher,* 27:620–625, 1989.

Frohlich, C. "Do Springboard Divers Violate Angular Momentum Conservation?" *American Journal of Physics,* 47:583–592, 1979.

Frohlich, C. "The Physics of Somersaulting and Twisting." *Scientific American,* 242:154–165, 1980.

Gaal, R. "Cat Tongues Are the Ultimate Detanglers." *APS News,* 26(1), 2017. Online.

Galileo. *Dialogue Concerning the Two Chief World Systems.* Translated by Stillman Drake. University of California Press, 1953.

Galli, J. R. "Angular Momentum Conservation and the Cat Twist." *Physics Teacher,* 33:404–407, 1995.

Gambale, F. "Does a Cat Always Land on Its Feet?" *Annals of Improbable Research,* 4:19, 1998.

Gardner, M. *My Best Mathematical and Logic Puzzles.* Dover Publications, New York, 1994.

Garnett, L. M. J. *The Women of Turkey and Their Folk-Lore.* D. Nutt, London, 1891.

Gauer, O., and H. Haber. *Man under Gravity-Free Conditions.* U.S. Government Printing Office, Washington, DC, 1950.

Gazenko, O. G., et al. "Harald von Beckh's Contribution to Aerospace Medicine Development (1917–1990)." *Acta Astronautica,* 43:43–45, 1998.

Ge, X.-S., and L.-Q. Chen. "Optimal Control of a Nonholonomic Motion Planning for a Free-Falling Cat." *Applied Mathematics and Mechanics,* 28:601–607, 2007.

Gerathewohl, S. J. "Comparative Studies on Animals and Human Subjects in the Gravity-Free State." *Journal of Aviation Medicine,* 25:412–419, 1954.

Gerathewohl, S. J., and H. D. Stallings. "The Labyrinthine Postural Reflex (Righting Reflex) in the Cat during Weightlessness." *Journal of Aviation Medicine,* 28:345–355, 1957.

Gihon, J. L. "Instantaneous Photography." *The Philadelphia Photographer,* 9:6–9, 1872.

Graves, R. P. *Life of Sir William Rowan Hamilton,* volume 3. Hodges, Figgis, Dublin, 1889.

Grayling, A. C. *Descartes.* Pocket Books, London, 2005.

Gross, R. S. "The Excitation of the Chandler Wobble." *Geophysical Research Letters,* 27:2329–2332, 2000.

Guyou, É. "Note relative à la communication de M. Marey." *Comptes Rendus,* 119:717–718, 1894.

Haber, F., and H. Haber. "Possible Methods of Producing the Gravity-Free State for Medical Research." *Journal of Aviation Medicine,* 21:395–400, 1950.

Haber, H. "The Human Body in Space." *Scientific American,* 184:16–19, 1951.

Hall, M. "On the Reflex Function of the Medulla Oblongata and Medulla Spinalis." *Philosophical Transactions of the Royal Society of London,* 123:635–665, 1833.

Haridy, R. "Boston Dynamics' Atlas Robot Can Now Chase You through the Woods," May 10, 2018. New Atlas website, https://newatlas.com/boston-dynamics-atlas-running/54573/.

Helios. "A New Sky Shade." *The Philadelphia Photographer,* 6:142–144, 1869.

Henry, J. P., E. R. Ballinger, P. J. Maher, and D. G. Simon. "Animal Studies of the Subgravity State during Rocket Flight." *Journal of Aviation Medicine,* 23:421–432, 1952.

Hetherington, J. H. "Letter to Ms. Lubkin," January 14, 1997. *Jack's Pages,* P. I. Engineering.com, http://xkeys.com/PIAboutUs/jacks/FDCWillard.php.

Holland, O. "The First Biologically Inspired Robots." *Robotica,* 21:351–363, 2003.

"A Horse's Motion Scientifically Determined." *Scientific American,* 39(16):241, 1878.

"How Do Cats Always Land on Their Feet?" March 31, 2016. Life in the Air, BBC One, available on YouTube at https://www.youtube.com/watch?v=sepYP_knGWc.

Hutton, C. *A Mathematical and Philosophical Dictionary,* volume 2. J. Johnson, London, 1795.

Iwai, T. "Classical and Quantum Mechanics of Jointed Rigid Bodies with Vanishing Total Angular Momentum." *Journal of Mathematical Physics,* 40:2381–2399, 1999.

Jusufi, A., D. I. Goldman, S. Revzen, and R. J. Full. "Active Tails Enhance Arboreal Acrobatics in Geckos." *Proceedings of the National Academy of Sciences,* 105:4215–4219, 2008.

Jusufi, A., D. T. Kawano, T. Libby, and R. J. Full. "Righting and Turning in Midair Using Appendage Inertia: Reptile Tails, Analytical Models and Bio-Inspired Robots." *Bioinspiration and Biomimetics,* 5:045001, 2010.

Jusufi, A., Y. Zeng, R. J. Full, and R. Dudley. "Aerial Righting Reflexes in Flightless Animals." *Integrative and Comparative Biology,* 51:937–943, 2011.

Kan, J., T. Z. Aziz, A. L. Green, and E. A. C. Pereira. "Biographical Sketch, Giles Brindley, FRS." *British Journal of Neurosurgery,* 28:704–706, 2014.

Kane, T. R., and M. P. Scher. "A Dynamical Explanation of the Falling Cat Phenomenon." *International Journal of Solids and Structures,* 5:663–670, 1969.

Kane, T. R., and M. P. Scher. "Human Self-Rotation by Means of Limb Movements." *Journal of Biomechanics,* 3:39–49, 1970.

Katz, J. I. "When Hot Water Freezes before Cold." *American Journal of Physics,* 77:27–29, 2009.

Kaufman, R. D. "The Electric Cat: Rotation without Net Overall Spin." *American Journal of Physics,* 81:147–152, 2013.

Kawamura, T. "Understanding of Falling Cat Phenomenon and Realization by Robot." *Journal of Robotics and Mechatronics,* 26:685–690, 2014.

Kennedy, H. C. *Peano: Life and Works of Giuseppe Peano.* D. Reidel, Dordrecht, 1980.

Kim, S., M. Spenko, S. Trujillo, B. Heyneman, V. Mattoli, and M. R. Cutkosky. "Whole Body Adhesion: Hierarchical, Directional and Distributed Control of Adhesive Forces for a Climbing Robot." In *IEEE International Conference on Robotics and Automation,* pages 1268–1273, 2007. Online.

Kulwicki, P. V., E. J. Schlei, and P. L. Vergamini. *Weightless Man: Self-Rotation Techniques.* Technical report AMRL-TDR-62–129. Aerospace Medical Research Laboratories, Wright-Patterson Air Force Base, OH, 1962.

Lankester, R. "The Problem of the Galloping Horse." In *Science from an Easy Chair,* pages 52–84. Henry Holt, New York, 1913.

Lasanta, A., F. V. Reyes, A. Prados, and A. Santos. "When the Hotter Cools More Quickly: Mpemba Effect in Granular Fluids." *Physical Review Letters,* 119:148001, 2017.

Lecornu, L. "Sur une application du principe des aires." *Comptes Rendus,* 119:899–900, 1894.

"Leopard Cub Falling Out of a Tree in the Serengeti NP, Tanzania," August 17, 2014. Zoom Safari Videos, available on YouTube at https://www.youtube.com/watch?v=m7iwnbkax-U.

Levenson, T. *Newton and the Counterfeiter.* Mariner Books, Boston, 2010.

Levine, D. N. "Sherrington's 'The Integrative Action of the Nervous System': A Centennial Appraisal." *Journal of Neuroscience,* 253:1–6, 2007.

Lévy, M. "Observations sur le principe des aires." *Comptes Rendus,* 119:718–721, 1894.

Libby, T., T. Y. Moore, E. Chang-Siu, D. Li, D. J. Coheren, A. Jusufi, and R. J. Full. "Tail-Assisted Pitch Control in Lizards, Robots and Dinosaurs." *Nature,* 481:181–184, 2012.

Liddell, E. G. T. "Charles Scott Sherrington, 1857–1952." *Obituary Notices of Fellows of the Royal Society,* 8:241–270, 1952.

Lu, Z., and O. Raz. "Nonequilibrium Thermodynamics of the Markovian Mpemba Effect and Its Inverse." *PNAS,* 114:5083–5088, 2017.

Magnus, O. *Rudolf Magnus—Physiologist and Pharmacologist.* Kluwer Academic Publishers, Dordrecht, 2002.

Magnus, R. "Animal Posture." *Proceedings of the Royal Society of London B,* 98:339–353, 1925.

Magnus, R. "Wie sich die fallende Katze in der Luft umdreht." *Archives néerlandaises de physiologie de l'homme et des animaux,* 7:218–222, 1922.

Malkin, Z., and N. Miller. "Chandler Wobble: Two More Large Phase Jumps Revealed." *Earth, Planets and Space,* 62:943–947, 2010.

Marey, É. J. *Animal Mechanism.* D. Appleton, New York, 1874.

Marey, É. J. "Des mouvements que certains animaux exécutent pour retomber sur leurs pieds, lorsquils sont précipités dun lieu élevé." *Comptes Rendus,* 119:714–717, 1894.

Marey, É. J. *La méthode graphique dans les sciences expérimentales et principalement en physiologie et en médecine.* G. Masson, Paris, 1885.

Marey, É. J. "Sur les allures du cheval reproduites par la photographie instantanée." *La Nature,* 1st semester:54, 1879.

Marsden, J., R. Montgomery, and T. Ratiu. "Reduction, Symmetry, and Phase in Mechanics." *Memoirs of the American Mathematical Society,* 88, 1990.

Maxwell, N. "Induction and Scientific Realism: Einstein versus van Fraassen Part Three: Einstein, Aim-Oriented Empiricism and the Discovery of Special and General Relativity." *British Journal for the Philosophy of Science,* 44:275–305, 1993.

McDonald, D. A. "How Does a Cat Fall on Its Feet?" *New Scientist,* 7:1647–1649, 1960.

McDonald, D. A. "How Does a Falling Cat Turn Over?" *St. Bartholomew's Hospital Journal,* 56:254–258, 1955.

McDonald, D. A. "How Does a Man Twist in the Air?" *New Scientist,* 10:501–503, 1961.

McDonald, D. A. "The Righting Movements of the Freely Falling Cat (Filmed at 1500 f.p.s.)." *Journal of Physiology—Paris,* 129:34–35, 1955.

Mead, C. A., and D. G. Truhlar. "On the Determination of Born-Oppenheimer Nuclear Motion Wave Functions Including Complications due to Conical Intersections and Identical Nuclei." *Journal of Chemical Physics,* 70:2284–2296, 1979.

Merbl, Y., J. Milgram, Y. Moed, U. Bibring, D. Peery, and I. Aroch. "Epidemiological, Clinical and Hematological Findings in Feline High Rise Syndrome in Israel: A Retrospective Case-Controlled Study of 107 Cats." *Israel Journal of Veterinary Medicine,* 68:28–37, 2013.

Montgomery, R. "Gauge Theory of the Falling Cat." *Fields Institute Communications,* 1:193–218, 1993.

Mpemba, E. B., and D. G. Osborne. "Cool?" *Physics Education,* 4:172–175, 1969.

Muller, H. R., and L. H. Weed. "Notes on the Falling Reflex of Cats." *American Journal of Physiology,* 40:373–379, 1916.

Muybridge, E. "Photographies instantanées des animaux en mouvement." *La Nature,* 1st semester:246, 1879.

Nadar, P. "Le nouveau president." *Paris Photographe,* 4:3–9, No. 1, 1894.

Newcomb, S. "On the Dynamics of the Earth's Rotation, with Respect to the Periodic Variations of Latitude." *Monthly Notices of the Royal Astronomical Society,* pages 336–341, 1892.

Newton, I. *The Mathematical Principles of Natural Philosophy.* Translated by Andrew Motte. H. D. Symonds, London, 1803.

Noel, A., and D. L. Hu. "Cats Use Hollow Papillae to Wick Saliva into Fur." *PNAS,* 115:12377–12382, 2018.

"Notes on Some Points Connected with the Progress of Astronomy during the Past Year." *Monthly Notices of the Royal Astronomical Society,* 53:295, 1893.

O'Leary, D. P., and M. J. Ravasio. "Simulation of Vestibular Semicircular

Canal Responses during Righting Movements of a Freely Falling Cat." *Biological Cybernetics,* 50:1–7, 1984.

Ouellette, J. "When Cold Warms Faster Than Hot." *Physics World,* December 2017.

Pancharatnam, S. "Generalized Theory of Interference, and Its Applications." *Proceedings of the Indian Academy of Sciences A,* 44:247, 1956.

Papazoglou, L. G., A. D. Galatos, M. N. Patsikas, I. Savas, L. Leontides, M. Trifonidou, and M. Karayianopoulou. "High-Rise Syndrome in Cats: 207 Cases (1988–1998)." *Australian Veterinary Practitioner,* 31:98–102, 2001.

"Par-ci, par-là." *La Joie de la Maison,* 202:706, 1894.

Parent, A. "Sur les corps qui nagent dans des liqueurs." *Histoire de l'Academie Royale,* pages 154–160, 1700.

Peano, G. "Il principio delle aree e la storia d'un gatto." *Rivista di Matematica,* 5:31–32, 1895.

Peano, G. "Sopra la spostamento del polo sulla terra." *Atti della Reale Accademia delle scienze di Torino,* 30:515–523, 1895.

Peano, G. "Sul moto del polo terrestre." *Atti dell'Accademia Nazionale dei Lincei,* 5:163–168, 1896.

Peano, G. "Sul moto del polo terrestre." *Atti della Reale Accademia delle scienze di Torino,* 30:845–852, 1895.

Peano, G. "Sul moto di un sistema nel quale sussistono moti interni variabili." *Atti dell'Accademia Nazionale dei Lincei,* 4:280–282, 1895.

Peano, G. "Sur une courbe, qui remplit route une aire plane." *Mathematische Annalen,* 36:157–160, 1890.

"Perpetual Motion." *Modern Medical Science (and the Sanitary Era),* 10:182, 1897.

"Philosophy and Common Sense." *Monthly Religious Magazine,* 29–30:298, 1863.

"Photographs of a Tumbling Cat." *Nature,* 51:80–81, 1849.

Pope, M. T., and G. Niemeyer. "Falling with Style: Sticking the Landing by Controlling Spin during Ballistic Flight." In *IEEE/RSJ International Conference on Intelligent Robots and Systems,* pages 3223–3230, 2017.

Potonniée, G. *The History of the Discovery of Photography.* Tennant and Ward, New York, 1936.

Putterman, E., and O. Raz. "The Square Cat." *American Journal of Physics,* 76:1040–1044, 2008.

Quitard, P. M. *Dictionnaire Étymologique, Historique et Anecdotique des Proverbes.* P. Pertrand, Paris, 1839.

Rademaker, G. G. J., and J. W. G. ter Braak. "Das Umdrehen der fallenden Katze in der Luft." *Acta Oto-Laryngologica,* 23:313–343, 1935.

Rankine, W. J. M. *Manual of Applied Mechanics.* Griffin, London, 1858.

Reis, P. M, S. Jung, J. M. Aristoff, and R. Stocker. "How Cats Lap: Water Uptake by *Felis catus.*" *Science,* 330:1231–1234, 2010.

Renner, E. *Pinhole Photography.* Focal Press, Boston, 2nd edition, 2000.

Robe, T. R., and T. R. Kane. "Dynamics of an Elastic Satellite—I." *International Journal of Solids and Structures,* 3:333–352, 1967.

Robinson, G. W. "The High Rise Trauma Syndrome in Cats." *Feline Practice,* 6:40–43, 1976.

Ross, C. H. *The Book of Cats.* Griffith and Farran, London, 1893.

Routh, E. J. *The Elementary Part of a Treatise on the Dynamics of a System of Rigid Bodies.* Macmillan, London, 1897.

Sadati, S. M. H., and A. Meghdari. "Singularity-Free Planning for a Robot Cat Freefall with Control Delay: Role of Limbs and Tail." In *8th International Conference on Mechanical and Aerospace Engineering,* pages 215–221, 2017.

Schock, G. J. D. "A Study of Animal Reflexes during Exposure to Subgravity and Weightlessness." *Aerospace Medicine,* 32:336–340, 1961.

Schrödinger, E. *The Present Situation in Quantum Mechanics.* Princeton University Press, Princeton, NJ, 1983. Translated reprint of original paper.

Seabrook, W. *Doctor Wood, Modern Wizard of the Laboratory.* Harcourt, Brace, New York, 1941.

Sherrington, C. S. *Inhibition as a Coordinative Factor.* Elsevier, Amsterdam, 1965.

Sherrington, C. S. *The Integrative Action of the Nervous System.* Charles Scribner's Sons, New York, 1906.

Sherrington, C. S. "Note on the Knee-Jerk and the Correlation of Action of Antagonistic Muscles." *Proceedings of the Royal Society of London,* 52:556–564, 1893.

Sherrington, C. S. "On Reciprocal Innervation of Antagonistic Muscles." Third note. *Proceedings of the Royal Society of London,* 60:414–417, 1896.

Shield, S., C. Fisher, and A. Patel. "A Spider-Inspired Dragline Enables Aerial Pitch Righting in a Mobile Robot." In *IEEE/RSJ International Conference on Intelligent Robots and Systems,* pages 319–324, 2015. Online.

Shields, B., W. S. P. Robertson, N. Redmond, R. Jobson, R. Visser, Z. Prime, and B. Cazzolato. "Falling Cat Robot Lands on Its Feet." In *Proceedings of Australasian Conference on Robotics and Automation, 2–4 Dec 2013,* 2013.

Skarda, E. "Cat Survives 19-Story Fall by Gliding Like a Flying Squirrel." *Time Magazine,* March 22, 2012. Online.

Smith, P. G., and T. R. Kane. "On the Dynamics of the Human Body in Free Fall." *Journal of Applied Mechanics,* 35:167–168, 1968.

Solnit, R. *River of Shadows.* Penguin Books, New York, 2003.

Stables, W. G. *"Cats": Their Points and Characteristics, with Curiosities of Cat Life, and a Chapter on Feline Ailments.* Dean and Son, London, 1874.

Stables, W. G. *From Ploughshare to Pulpit: A Tale of the Battle of Life.* James Nisbet, London, 1895.

"The Steam Man." *Scientific American,* 68:233, 1893.

Stepantsov, V., A. Yeremin, and S. Alekperov. "Maneuvering in Free Space." *NASA TT F-9883,* 1966.

Stokes, G. G. *Memoir and Scientific Correspondence.* Cambridge University Press, Cambridge, 1907.

Stratton, J. A. "Harold Eugene Edgerton (April 6, 1903–January 4, 1990)." *Proceedings of the American Philosophical Society,* 135:444–450, 1991.

Studnicka, F., J. Slegr, and D. Stegner. "Free Fall of a Cat—Freshman Physics Exercise." *European Journal of Physics,* 37:045002, 2016.

Tait, P. G. "Clerk-Maxwell's Scientific Work." *Nature,* 21:317–321, 1880.

Takagi, S., M. Arahori, H. Chijiiwa, M. Tsuzuki, Y. Hataji, and K. Fujita.

"There's No Ball without Noise: Cats' Prediction of an Object from Noise." *Animal Cognition,* 19:1043–1047, 2016.

Teitel, A. S. *Breaking the Chains of Gravity.* Bloomsbury Sigma, New York, 2016.

Tesla: Master of Lightning, PBS, website materials on Life and Legacy, http://www.pbs.org/tesla/ll/story_youth.html.

Thompson, G. "Spiders and the Electric Light." *Science,* 9:92, 1887.

Thone, F. "Right Side Up." *Science News-Letter,* 25:90–91, 1934.

Tomson, G. R. *Concerning Cats: A Book of Poems by Many Authors.* Frederick A. Stokes, New York, 1892.

Toulouse, E. "Nécrologie—Marey." *Revue Scientifique,* 5:673–675, T. 1 1904.

Triantafyllou, M. S., and G. S. Triantafyllou. "An Efficient Swimming Machine." *Scientific American,* 272:64–70, March 1995.

Vandiver, J. K., and P. Kennedy. "Harold Eugene Edgerton (1903–1990)." *Biographical Memoirs,* 86:1–23, 2005.

Verbiest, H. "In Memoriam Prof. Dr. G. G. J. Rademaker." *Nederlands Tijdschrift voor Geneeskunde,* 101:849–851, 1957.

Vincent, J. F. V., O. A. Bogatyreva, N. R. Bogatyrev, A. Bowyer, and A-K. Pahl. "Biomimetics: Its Practice and Theory." *Journal of the Royal Society Interface,* 3:471–482, 2006.

Vnuk, D., B. Pirkic, D. Maticic, B. Radisic, M. Stejskal, T. Babic, M. Kreszinger, and N. Lemo. "Feline High-Rise Syndrome: 119 Cases (1998–2001)." *Journal of Feline Medicine and Surgery,* 6:305–312, 2004.

Volterra, V. "Il Presidente Brioschi dà comunicazione della seguente lettera, ricevuta dal Corrispondente V. Volterra." *Atti dell'Accademia Nazionale dei Lincei,* 5:4–7, 1896.

Volterra, V. "Osservazioni sulla mia Nota: 'Sui moti periodici del polo terrestre.'" *Atti della Reale Accademia delle scienze di Torino,* 30:817–820, 1895.

Volterra, V. "Sui moti periodici del polo terrestre." *Atti della Reale Accademia delle scienze di Torino,* 30:547–561, 1895.

Volterra, V. "Sulla rotazione di un corpo in cui esistono sistemi ciclici." *Atti dell'Accademia Nazionale dei Lincei,* 4:93–97, 1895.

Volterra, V. "Sulla teoria dei moti del polo nella ipotesi della plasticità terrestre." *Atti della Reale Accademia delle scienze di Torino,* 30:729–743, 1895.

Volterra, V. "Sulla teoria dei moti del polo terrestre." *Atti della Reale Accademia delle scienze di Torino,* 30:301–306, 1895.

Volterra, V. "Sulla teoria dei movimenti del polo terrestre." *Astronomische Nachrichten,* 138:33–52, 1895.

Volterra, V. "Sul moto di un sistema nel quale sussistono moti interni variabili." *Atti dell'Accademia Nazionale dei Lincei,* 4:107–110, 1895.

von Beckh, H. J. A. "Experiments with Animals and Human Subjects under Sub and Zero-Gravity Conditions during the Dive and Parabolic Flight." *Journal of Aviation Medicine,* 25:235–241, 1954.

Wagstaff, K. "Purr-plexed? Cats Teach a Robot How to Land on Its Feet." *Today,* October 14, 2016. Online.

Walker, C., C. J. Vierck Jr., and L. A. Ritz. "Balance in the Cat: Role of the Tail and Effects of Sacrocaudal Transection." *Behavioural Brain Research,* 91:41–47, 1998.

Walter, W. G. "An Imitation of Life." *Scientific American,* pages 42–45, May 1950.

Weber, R. L. *More Random Walks in Science.* Taylor and Francis, New York, 1982.

Weed, L. H. "Observations upon Decerebrate Rigidity." *Journal of Physiology,* 48:205–227, 1914.

Wehrey, C. "Hubble and Copernicus," November 8, 2012. *Verso: The Blog of the Huntington Library, Art Collections, and Botanical Gardens,* http://huntingtonblogs.org/2012/11/hubble-and-copernicus/.

Whitney, W. O., and C. J. Mehlhaff. "High-Rise Syndrome in Cats." *Journal of the American Veterinary Medical Association,* 191:1399–1403, 1987.

Whitsett, C. E., Jr. *Some Dynamic Response Characteristics of Weightless Man.* Technical Report AMRL-TDR-63–18. Aerospace Medical Research Laboratories, Wright-Patterson Air Force Base, OH, 1963.

Whittaker, C. "Vito Volterra. 1860–1940." *Obituary Notices of Fellows of the Royal Society,* 3:691–729, 1941.

"Why Cats Always Land on Their Feet." *Current Opinion,* 17:42, 1895.

Wojciechowski, B., I. Owczarek, and G. Bednarz. "Freezing of Aqueous Solutions Containing Gases." *Crystal Research and Technology,* 23:843–848, 1988.

Woodruff, T. O. "WoodruffLetter" (Letter to Jack Hetherington), November 26, 1975. *Jack's Pages,* P. I. Engineering.com, http://xkeys.com/PIAboutUs/jacks/FDCWillard.php.

Woods, G. S. "Stables, William Gordon." In *Oxford Dictionary of National Biography.* Oxford University Press, Oxford, 2004.

The World of Wonders. Cassell, London, 1891.

Wright, J. M. *Alma Mater; or, Seven Years at the University of Cambridge.* Black, Young and Young, London, 1827.

Wright, W. *Flying,* pages 87–94, March 1902.

Yamafuji, K., T. Kobayashi, and T. Kawamura. "Elucidation of Twisting Motion of a Falling Cat and Its Realization by a Robot." *Journal of the Robotics Society of Japan*, 10:648–654, 1992.

Yeadon, M. R. "The Biomechanics of Twisting Somersaults. Part III: Aerial Twist." *Journal of Sports Science,* 11:209–218, 1993.

Zhao, J., L. Li, and B. Feng. "Effect of Swing Legs on Turning Motion of a Freefalling Cat Robot." In *Proceedings of 2017 IEEE International Conference on Mechatronics and Automation,* pages 658–664, 2017.

Zhen, S., K. Huang, H. Zhao, and Y-H. Chen. "Why Can a Free-Falling Cat Always Manage to Land Safely on Its Feet?" *Nonlinear Dynamics,* 79:2237–2250, 2015.

Acknowledgments

A book of this nature takes a lot of effort, both intellectually and emotionally. I have been aided by a large number of friends, colleagues, and other generous souls and would like to take a few final moments to express my appreciation.

First, my greatest thanks go to Sarah Addy, my friend and a talented artist who drew all of the beautiful illustrations of cats that were beyond my capability to draw. It is because of her that you aren't seeing stick figures in improbable poses throughout the book.

Many of the historical papers that I discuss were not written in English. For the most part, I relied on Google Translate to get me through them—fortunately, scientific papers are usually written in a dry and to-the-point manner, making the interpretations straightforward. A few key passages in French required more elegant translations, and I would like to thank my friend Jana Sloan van Geest for providing them. Thanks also to Dr. Jens Foell for translating a key quotation by Albert Einstein from the German. I would also like to thank my longtime friend Rebecca Starkey for helpful advice on negotiating the library world.

I reached out to many scientists for interviews or information during the writing of the book; unfortunately, I received far fewer answers than requests (I suspect a lot of people couldn't believe that this was a serious project). So I am especially grateful to Alexis Noel of Georgia Tech, Will Robertson of the University of Adelaide, Jack Hetherington of Michigan State University and P. I. Engineering and his wife Marge, and Michael Berry of the University of Bristol for generously taking the time to answer questions. (Any inaccuracies that might appear in the book, however, I take full credit for.)

My friends have helped keep my morale up during the long, often stressful process of writing. I would like to give special thanks to Beth Szabo, Mahy El-Kouedi, and Kayla Arenas for being dear friends. I would also like to thank (as I do in every book) my skating coach, Tappie Dellinger; my guitar instructor, Toby Watson; and my skydiving friends at Skydive Carolina for keeping me distracted and entertained. Thanks, as always, to my parents, Pat Gbur and John Gbur, for everything.

I had to acquire lots of figure permissions for this book and appreciate all the assistance I received in the process. I would like to give a special shoutout to Pam Day, managing editor of *Aerospace Medicine and Human Performance*, for not only providing permissions but offering the alliterative phrase "Falling felines are a fascinating phenomenon."

Figures cost money to reproduce in a book, and some of them cost an incredible amount. Near the end of 2018, when I launched a GoFundMe to ask for funding help with the permissions, my friends, online and off, came through to a degree for which I will forever be grateful. Special shoutouts and acknowledgments are owed to the following people (and their pets): Mark Mancini; Brian, Brennan, and Taz Cox; Yongtao Zhang; the Bigelow CAT group of the University of Rochester; Laura Kinnischtzke and her dear pets, Bert (coolest Maine coon ever) and Dixie (best beagle friend); Karteek and Maitreyi on behalf of Futurecat Kuppuswamy; Azure Hansen, Ziggy, and Anastasia; Dave Curtis and Jack the Dog; Ronald A. Ambrose

Jr. and his cat, Mr. Skittles; Damon Diehl and Brad Craddock, in memory of Elphaba and Erasmus; Chance (Buddy) Cruickshank; Jeff Sensabaugh, in memory of his cat Nutmeg; Sue, Bowzer, and (the late) Sera; Gareth Dew; John Gbur (aka Dad); Rebecca Stefoff and Xerxes; Steve Carrow; Bryan R. Gibson; Lawrence Rogers; Jen Cross and Zool; Aaron and Sara Golas and Link the cat; Thomas Swanson; Chris Souza on behalf of Skippy, the once-feral cat who adopted them, and Pikka-Charr, the chinchilla totally worth the two-hundred-mile rescue trip; Michele Banks and Teapot; Joanne Power and Mo; Marian I. Tjaden and Jet; Jason Thalken; Steve Cooke. An extra-special thanks goes to Pascale Lane and Dottie and to Jana Middleton!

Finally, I would like to thank Joseph Calamia and Mary Pasti, my editors at Yale University Press, for their help in getting this book published and making it the best it can be.

Index

Page numbers in *italics* indicate figures.